AVIAN MONOGAMY

ORNITHOLOGICAL MONOGRAPHS

This series, published by the American Ornithologists' Union, has been established for major papers too long for inclusion in the Union's journal, *The Auk.* Publication has been made possible through the generosity of the late Mrs. Carll Tucker and the Marcia Brady Tucker Foundation, Inc.

Correspondence concerning manuscripts for publication in the series should be addressed to the Editor, Dr. David W. Johnston, Department of Biology, George Mason University, Fairfax, VA 22030.

Copies of *Ornithological Monographs* may be ordered from the Assistant to the Treasurer of the AOU, Frank R. Moore, Department of Biology, University of Southern Mississippi, Southern Station Box 5018, Hattiesburg, Mississippi 39406. (See price list on back and inside back covers.)

Ornithological Monographs, No. 37, vi + 121 pp.

Editors of *Ornithological Monographs,* Mercedes S. Foster and David W. Johnston

Special Reviewers for this issue, Walter D. Koenig, Hastings Reservation, Star Route Box 80, Carmel Valley, CA 93924; Lewis W. Oring, Department of Biology, Box 8238, University Station, Grand Forks, ND 58202

Authors, Patricia Adair Gowaty, Department of Biological Sciences, Clemson University, Clemson, SC 29631; Douglas W. Mock, Department of Zoology, University of Oklahoma, Norman, OK 73019

First received, 23 August 1983; accepted 29 February 1984; final revision completed 8 October 1984

Issued October 17, 1985

Price $11.00 prepaid ($9.00 to AOU members).

Library of Congress Catalogue Card Number 85-647080

Printed by the Allen Press, Inc., Lawrence, Kansas 66044

ISBN: 0-943610-45-1

AVIAN MONOGAMY

EDITED BY

PATRICIA ADAIR GOWATY

AND

DOUGLAS W. MOCK
Department of Zoology
University of Oklahoma
Norman, Oklahoma 73019

ORNITHOLOGICAL MONOGRAPHS NO. 37

PUBLISHED BY
THE AMERICAN ORNITHOLOGISTS' UNION
WASHINGTON, D.C.
1985

TABLE OF CONTENTS

PREFACE

The symposium, Avian Monogamy, was presented at the 100th Stated Meeting of the American Ornithologists' Union at the Field Museum, Chicago, Illinois in October 1982. The idea for a symposium arose after it occurred to us that avian monogamy lacked general attention from the ornithological community. We invited speakers from among the few we knew who at the time were actively interested in theoretical and empirical aspects of monogamy among birds. The morning symposium was followed that afternoon by a related contributed paper session also on monogamy. Contributors to this volume spoke in one of those sessions.

Patricia Adair Gowaty
Douglas W. Mock

CHAPTER 1

AN INTRODUCTION TO THE NEGLECTED MATING SYSTEM

Douglas W. Mock

Department of Zoology, University of Oklahoma, Norman, OK 73019

An amusing paradox has developed over the past two decades: the species doing the research on evolutionary aspects of mating systems generally regards itself as monogamous while devoting the great bulk of its scientific effort toward elucidating the principles underlying polygamy. Admittedly, skepticism is justified about how monogamous humans really are, but the neglect of scientific inquiry into the causes of monogamy remains surprising. This is especially interesting in ornithology for two reasons: monogamy is the predominant mating system in birds and the biological interest in mating systems has been fundamentally shaped by ornithologists (e.g., David Lack, John Crook, Frank Pitelka, Gordon Orians, Jerram Brown, and many others). Whether the fraction of birds that are primarily monogamous is 91% (as estimated by Lack 1968) or somewhat less, monogamy's prevalence among birds requires explanation. Too little is known of why it has been maintained in so many avian populations while being rare in virtually all other taxa.

Perhaps the neglect of monogamy is simply an artifact of the research protocol that rewards pursuit of the extreme first. In the area of mating systems, most light has been shed on sexual selection theory by analyzing how the critical component, intrasexual variance in mating success, reaches its maximum. Quantifications of male and female variance, both in the lab (e.g., Bateman 1948) and field (e.g., LeBoeuf 1974; Clutton-Brock et al. 1982), have spawned important insights into how that variance is created. By contrast, mating variance in monogamous species is expected to be relatively low, presumably reflected in the subdued or even drab secondary sexual characters of many participants. Monogamous birds do not establish spectacular leks and only occasionally are highly ornamented. On the surface, monogamy has seemed relatively tame and uniform, with a single male mating routinely with a single female. Not only has sexual selection appeared feeble, but the whole package seems bland.

The primary objective of this volume is to penetrate below the surface of monogamy in general and avian monogamy in particular. The progress made with non-monogamous mating systems can be used to steer investigations of monogamy, but fresh approaches are also in use. These papers are intended to precipitate new interest in the subtle machinations of sexual selection in monogamous birds and to stimulate thinking about long-standing problems such as the many cases of marked sexual dimorphism existing in monogamous species.

Sexual Selection and Monogamy

From the original formulation of sexual selection theory (Darwin 1871) and its recent renaissance (e.g., Campbell 1972; Emlen and Oring 1977; Maynard Smith 1978; Blum and Blum 1979; Dunbar 1982), it has been clear that sexual selection can and does operate in monogamous animals, but little attention has been given to its precise pathways and intensity. Darwin (1871) suggested that early mating could confer fitness benefits if females that were ready to breed first subsequently attained the greatest reproductive output; males obtaining such mates would enjoy disproportionately high success also. This idea was refined (Fisher 1930) and tested with both field data and modeling (O'Donald 1974) to clarify the components most likely shaped by selection. Few other leads have been followed.

Sexual selection operates through the processes of intrasexual selection (competition among members of the limited sex for access to mates of the limiting sex: usually males competing for females) and intersexual, or "epigamic," selection (mating preferences by members of the opposite sex). Intrasexual competition can be conducted in a variety of ways, including (but not restricted to) overt combat. The intersexual component is probably very complex, especially in monogamy, where both sexes can gain by being "choosy." Together, these processes can produce variable degrees of within-sex variance in mating success, the key empirical measure of sexual selection's overall intensity (Bateman 1948).

Understanding the diversity of monogamous mating systems therefore hinges on identifying sources of intrasexual variance in mating success under the apparent social confines of monogamy. Such variance can derive from several sources (even within the restriction that each breeding adult be limited to one primary mate), including the possibilities of having minor extra-pair liaisons (i.e., an arrangement intermediate between total fidelity and bigamy), of having no mate whatever, and of having a mate with relatively high or low reproductive value. For example, it has been known for many years that territorial behavior can disenfranchise a sizeable fraction of the potential breeding population, referred to collectively and vaguely as "floaters." Successful members of each sex socially exclude others from breeding, thus establishing variance in mating success and the potential for selection favoring phenotypic characters that confer that success (e.g., features that enable acquisition of a territory). Experimental removal of breeders has demonstrated that these "floaters" are ready and willing to breed (Brown 1969). Unfortunately, it is logistically very difficult to assess the size of the floater population (most bird census techniques rely on the behavioral conspicuousness of the successful territory holders), so we have no quantitative index of the impact of territorial exclusion on the intensity of sexual selection. Qualitatively, however, this kind of exclusion is comparable to the more spectacular forms of mating exclusion found in some polygynous species (e.g., intimidating, evicting, or even killing competitors). From the genetic standpoint, socially imposed celibacy differs from sterility or death only by its impermanence.

Animals generally affect mating success variance in two ways: they directly promote their own success and/or they depress the success of others, thereby gaining a relative advantage. Territorial exclusion falls under both categories, because the resident simultaneously assures itself of resources for raising progeny while denying that advantage to floaters (Verner 1977). Even more dramatically, the success of conspecifics can be depressed via infanticide (including egg-destruction, Picman 1977). Infanticide by adult birds is probably much more common than is generally appreciated (reviewed by Mock 1984), particularly in non-monogamous species, which lack biparental defense. Its impact on reproductive success in monogamous species is not well understood, but may be important in colonial species (Mock 1984), communal/cooperative breeders (Vehrencamp 1977; Trail et al. 1981; Mumme et al. 1983b; Stacey and Edwards 1983), and perhaps even in "typical" territorial species (e.g., Yom-Tov 1974).

Alternatively, variance in mating success can arise when individuals manage to mate successfully outside the primary "bond," even while making substantial contributions of parental care. Dubbed "mixed reproductive strategies" by Trivers (1972), this has been reported for males in many species of apparently monogamous birds (see reviews by McKinney et al. 1983; McKinney, Chap. 6), forcing a reconsideration of the latitude allowed the term "monogamy." Recent treatments of intraspecific brood parasitism (Andersson and Eriksson 1982; Gowaty, Chap. 2) have argued that such behavior is the genetic equivalent of more familiar male philandering, because it conforms to Payne's (1977) characterization of brood parasitism as "theft of parental investment." However achieved, usurpation of a competitor's reproductive success contributes to within-sex variance and hence to the potency of sexual selection.

The importance of mate choice (epigamic selection) has proven more difficult to assess. Are male Northern Cardinals (*Cardinalis cardinalis*) red because that hue repels male competitors, because it attracts females, or both? Burley (1977a, 1981a, Chap. 3) has taken original and provocative experimental approaches to the issue of mate choice, a subject that has largely defied field study.

Puzzles of Monogamy

Monogamous mating is highly correlated with relatively large contributions of postnatal parental investment (PI), from both males and females (Lack 1968). Because anisogamy (sexual size dimorphism of gametes) is theoretically responsible for the usual pattern of minimal male contributions (Trivers 1972; Parker et al. 1972), the secondary development of large male PI is intriguing. Monogamy can evolve only when this phylogenetic inertia has been overcome, so the factors responsible for such a change are well worth consideration.

Among birds, male contribution to PI varies greatly, with males of many polygynous species providing only gametes and males of polyandrous species providing nearly everything but ova. Less appreciated, but no less interesting, is the fact that a huge range of male PI contributions occurs among monogamous bird species also. In a few species (e.g., Willow Ptarmigan, *Lagopus lagopus*) the male provides only sperm and some vigilance, but females seem too dispersed (and/or too aggressive?) to permit bigamy (Hannon 1984); this would seem to be a variation of "facultative monogamy" (Kleiman 1977). In Eastern Bluebirds (*Sialia sialis*), males contribute both sperm and the nesting cavity, but are not essential for successful brood-rearing (Gowaty 1983); this type of facultative monogamy seems to hinge on the overdispersion of nest cavities. In many "typical" monogamous species, the male defends a territory in which the female collects food for the brood. In still other species, he provides some of the food directly, but may not incubate (e.g., Barn Swallows, *Hirundo rustica*), whereas in many more species (e.g., herons), he provides both food and incubation. Finally, males of many species (e.g., gulls, terns, swans, and storks) take substantial personal risks to protect their offspring from predators. Given that male mammals seldom provide more than genes (monogamy is believed to be the primary mating system in only 3% of mammalian species: Kleiman 1977), the wealth of comparative material in the Class Aves offers an obvious key toward discovering the ecological correlates of a large male PI.

By now it is clear that monogamy is not *a* mating system, but a diverse array of reproductive strategies that may have so little in common as to defy unified definition (see Gowaty, Chap. 2, for definitions). Thus, a second goal of this collection is to draw attention to how our language sacrifices information by lumping so much variation under one heading; the diversity merits study. In practice, monogamy has served as a catch-all, where species are assigned only when they fail to satisfy the more easily specified criteria for polygyny or polyandry. It has become a mating-system-by-default. For example, in their review of passerine mating systems, Verner and Willson (1969) assigned species to monogamy only if 95% of the studied mating units were neither polygynous nor polyandrous. Similarly, attempts to establish even a qualitative theoretical base for monogamy have simply reversed the logic of polygyny models (Wittenberger and Tilson 1980; Gowaty 1981a). No quantitative models have been produced yet for the evolution of monogamy *per se.*

General Factors Favoring the Evolution of Monogamy

Because anisogamy is an extremely primitive character among sexual organisms, presumably predating the evolution of complex parental care, it is logical to assume that monogamy was not the primitive mating system; indeed, it has surely evolved independently many times. Though the phylogeny of monogamy is not our primary concern here, various aspects of phylogenetic inertia (Wilson 1975) and ecology can be identified as likely contributing factors to the evolution of avian monogamy, including: (1) parental care needs of the young, (2) oviparity, (3) population structure, (4) spatio-temporal dispersion of critical resources, and (5) lack of specializations for uniparental care (e.g., no lactation).

Monogamy is commonly associated with "K-strategy" ontogenetic patterns, where the postnatal needs of offspring may outstrip a single parent's ability to provide. Obviously, this imbalance is related to the kinds and distribution of resources used in parenting. When uniparental care is sufficient to meet the brood's requirements, desertion by one ("emancipated") parent is often more profitable than staying with the brood (Trivers 1972; Maynard Smith 1977). Conversely, whenever the combined efforts of two parents allow considerably greater offspring survivorship or quality, continued investment may become the better option.

Unless accompanied by synchronized laying and/or sperm storage (as in most oviparous reptiles, amphibians, fish, and insects), oviparity reduces the likelihood that the male parent can gain from desertion. In contrast with viviparity, it encumbers the female for a much smaller fraction of the time for embryonic development (in birds, often just a day), effectively forcing the male to remain and continue copulating (and in many cases mate-guarding) at least until the last ovum is fertilized. Insofar as this extended period of male commitment can substantially reduce his reproductive alternatives, bird-type oviparity increases the chances for monogamous brood-rearing.

Population structure and breeding synchrony are expected to influence the evolution of monogamy through the deserting parent's probability of finding additional mates (Maynard Smith 1977). Similarly, age structure of the population theoretically influences the operational sex-ratio (number of fertilizable females to the number of sexually active males: Emlen and Oring 1977) and thus constrains non-monogamous opportunities. By pointing out how life history parameters affect lifetime breeding options, Murray (Chap. 8) calls attention to frequently ignored demographic aspects of monogamous reproductive strategies.

The effects of ecological resources on mating systems has been a central topic in the mating systems literature (e.g., Orians 1969; Pitelka et al. 1974; Emlen and Oring 1977, Wittenberger 1979, 1981; Oring 1982) and has been considered specifically with regard to monogamy (Wittenberger and Tilson 1980). Generally, critical resources must not be so readily able to be monopolized as to force females into accepting little or no PI contributions from the resource-controlling males. Resource dispersion may promote facultative monogamy if the resulting population density is so low that single males encounter females rarely during the breeding period. Because of the avian potential for extreme mobility, such monogamy is probably relatively rare in birds.

Some Future Research Directions

Here I want to sketch out six general categories of questions about evolutionary aspects of monogamy, some of which can be and are being addressed and some of which must await inspired moments of research cleverness to become feasible.

Male investment.—If the central riddle of monogamy is, indeed, why males contribute so much parental care, then we clearly need to know more about the nature of that effort. In particular, it would be instructive to know its value to male fitness (benefits) relative to the presumed sacrifice of not pursuing additional mates (costs). First, is male PI essential, as predicted? If biparental care is required, male options are constrained far more severely than otherwise. A few studies have begun testing this assumption experimentally, by removing males at various points in the breeding cycle and measuring the impact of that loss on brood survival. Provocatively, this has shown that monogamous males are not always essential (e.g., Gowaty 1983; Hannon 1984) for brood-rearing. (It remains possible, of course, that such males do provide something of value to the female that may not be revealed in the current brood's success, but why quibble?) This seems to be a promising approach for understanding the degree to which monogamous males are indispensable vs emancipated, thereby indicating the relative importance of this vs other constraints. Such experiments on a variety of monogamous species would be most interesting because such a literature would allow comparative testing of the many male desertion hypotheses (e.g., Trivers 1972; Maynard Smith 1977; Gladstone 1979) and improve our understanding of the long-term consequences of forced uniparental care. In many species, it would be equally interesting to remove females and explore the male's capacity for uniparental care as well. Alternatively, the ecological factors that demand postzygotic male investment can be explored through direct comparisons when closely related uniparental and biparental species breed sympatrically (e.g., Post and Greenlaw 1982; Beehler, Chap. 7; McKinney, Chap. 6).

Population structure.—A general need exists for accurate methods of assessing the composition of whole local populations and discerning which bird gets to breed and which does not. This will be a major logistic challenge because many of the birds whose activities and existence we need to know about are under great social pressure from dominant conspecifics to be inconspicuous. Nevertheless, accurate measurement of floater populations and operational sex ratios would be extremely interesting and instructive (e.g., Payne 1979; Gowaty, Chap. 2), both for assessing the intensity of sexual selection and for understanding the reproductive alternatives available to each sex (Maynard Smith 1977, 1978).

Sexual selection processes.—The search for sources of within-sex variance in reproductive success of monogamous animals clearly needs to be extended. Some of the features of intrasexual selection, for example, that have been studied well in their more exaggerated polygynous forms should be scrutinized in monogamous birds. Such behavioral events as overt combat, nest sabotage, and egg destruction may be rare but important contributors to variance. Hopefully, persons working on various other aspects of avian breeding biology will be alert for such possibilities. Beyond the anecdotal level, it seems likely that serendipitous field students will discover unexpected and highly complex mixtures of intrasexual and intersexual selection and concentrate on such systems. For example, Fujioka and Yamagishi (1981) reported that neighboring male Cattle Egrets (*Bubulcus ibis*) have dominance relationships among themselves that confer advantages in extra-pair copulations with each others' mates. Although the genetic impact of this behavior is not known, it is provocative and may occur in other monogamous, colonial species as well (e.g., White Ibis, *Eudocimus albus*; Kushlan 1973; Rudegeair 1975).

Intersexual "mate choice" preferences have proven to be less easily studied in the wild, although casual observations of behavioral selectivity are worth reporting (e.g., Mock 1979). In monogamous mating systems, reciprocal choosiness would be favored (Burley 1977a, Chap. 3; Bluhm, Chap. 4), because both sexes invest heavily and rely on each other. Laboratory experimentation seems to hold the most promise at present because of difficulties in measuring "choosiness" in the wild (Payne 1979). To my knowledge, no quantitative measures are known for the impact that mate choice has on within-sex variance in reproductive success of wild animals.

The potential importance of mate-choice to the evolution and operation of monogamy is enormous, extending beyond the acquisition of primary mates to the realm of extra-pair matings as well. It is not clear, for example, the degree to which so-called "forced copulations" (FC) are, in fact, contrary to the female's interests and wishes in some species (Thornhill 1980b), nor how (whether?) such males have the proximate power to effect FCs without female cooperation (Lumpkin 1981).

Offspring requirements.—The ecological factors believed to control how much investment progeny need are many and complex. However, because these requirements directly affect the fundamental issue of biparental care and its division, monogamy will not be understood thoroughly until they are. Growth requirements, ontogenetic patterns, predation pressure, thermoregulation, and nutrition circumscribe parental options and thus affect the type of mating system. Even within populations there must be unknown amounts of variability in the division of PI between the individual mates, which (theoretically) might benefit from selfishly extracting more than half of the offsprings' needs from the parental partner. Whether any such patterns of mutual exploitation actually occur in monogamous birds is unknown. One conceptual problem that remains unsolved in all studies of PI is the lack of a common currency by which all contributions are measured. Time and energy budgets have been estimated for many avian activities and a vague assumption exists that time and calories contributed to offspring are interchangeable at least in principal, but other types of PI seem less easily converted. Perhaps the most obvious of these is risk, the chance of being killed while, say, trying to deter predators. Obviously, total PI will be harder to quantify in species where parents take such risks than in species where parents eschew all dangerous confrontations.

Genetic relatedness.—At the very heart of all reproductive success measurements lies the assumption that observed mating patterns translate directly into gametic success, yet challenges are growing to that assumption (e.g., see reviews by McKinney et al. 1983; Gowaty, Chap. 2). Routine and accurate assessment of paternity, maternity, and sibling relatedness may be the largest single logistic obstacle to our study of sexual selection in the field (Sherman 1981; Mock 1983) and is likely to produce a flurry of counter-intuitive discoveries as new techniques are applied (e.g., Bray et al. 1975; Gowaty and Karlin 1984).

Extensions to family structure.—Finally, I am convinced that the so-called "nuclear" monogamous family is a highly attractive system for studying the interplay of genetic and ecological conflicts. The monogamous family can be viewed as a social microcosm with three dimensions: between mates (the "pair-bond"), between parents and offspring, and between siblings. Under conditions of outbreeding, the parents can be expected to share few genes through common descent and, therefore, to gain little through nepotistic generosity towards each other. In long-lived species where pair-bonds are brief (e.g., single season or less), relatively few constraints should exist on mutual exploitation ("sexual conflict" of Parker 1979; see also Gowaty, Chap. 2), which may pay dividends to the individual in terms of future reproductive success with other partners (i.e., increased longevity). By contrast, in species showing patterns of long-term pair-bonds (e.g., swans and geese), these dividends should be reduced because the individual's reproductive future depends substantially on its partner's welfare.

The other social dimensions of nuclear family structure, parent-offspring and intersibling, feature very high coefficients of relatedness ($r = 0.5$, assuming parental mate fidelity). Many circumstances are known in which this commonality of genetic interests is opposed by resource shortages, forcing overt conflict among the participants. Nestlings of many species behave in spectacularly selfish ways at such times, frequently causing the death or developmental retardation of siblings (reviewed by O'Connor 1978; Stinson 1979; Mock 1984). Parents play a variety of important roles in such brood reduction systems, ranging from overt infanticide of some or all young to more subtle manipulations of the brood-members' competitive abilities (Ricklefs 1965; Alexander 1974; O'Connor 1978; Hahn 1981; Mock 1984). These systems have been proposed as hotbeds of Trivers' (1974) concept of parent-offspring conflict (O'Connor 1978), although this possibility remains largely unexplored.

Finally, although I have emphasized the curious aspects of male PI and alternative mating options in this discussion, it is certain that increased research scrutiny of monogamy will uncover new facets of female strategies as well (e.g., Wasser 1983). Our ignorance of female variations on the monogamy theme is almost complete and thus offers many research opportunities for the future.

The Component Chapters

This volume is not intended as an exhaustive review of all topics relevant to the study of avian monogamy; rather it is a representative sample of many of the issues and approaches currently being used in the early stages of such study. In the process, several areas are reviewed in detail, including the problems associated with measuring within-sex variance in reproductive success (Gowaty), demographic constraints on mating system evolution (Murray), coevolutionary aspects of epigamic selection in monogamy (Burley), and the diversity of monogamy within a single taxon (McKinney).

In addition, three major empirical approaches are showcased. Modern long-term field observations of single-species (Anderson) or multi-species populations (Beehler), which necessarily rely on correlational analyses to indicate probable causal factors, have always been the backbone of ornithological studies in behavioral ecology. Such projects concern animals in their natural ecological contexts and provide the natural history foundations on which experimental refinements must build.

The second category of approach in evidence here is experimentation. Both the contributions of Burley and Bluhm involve the use of captive birds to assay the mechanism and importance of mate-choice phenomena in monogamous birds. The combination of Bluhm's experiment and Anderson's field data is particularly appealing because both concern the same species, Canvasbacks (*Aythya valisineria*), from the same Canadian population.

Thirdly, the analysis of genetic relatedness has led to specialized techniques for identifying kin (paternity and maternity) via protein electrophoresis (Gowaty) and phenotypic markers (McKinney).

Not surprisingly, this pluralism of scientific approaches has produced a varied collection, full of exciting and unavoidably tentative ideas. In the opening chapter, Patricia Adair Gowaty focuses attention on the likelihood that what we observe as monogamy in the field (male-female consorts) may or may not reflect the true genetic parentage of the offspring. Her data on the mixed parentage of Eastern Bluebirds (Gowaty and Karlin 1984) have documented both the predicted vulnerability of males (e.g., Trivers 1972; Alexander 1974) and a less-expected similar dilemma for females. If such patterns turn out to be commonplace, substantial problems arise for the usual assumption of congruence between "apparent" monogamy and the genetic results of monogamy.

Next, Nancy Burley addresses the elusive sexual selection process of mate-choice, with particular attention to the role played by non-functional or "aesthetic" phenotypic characteristics. The general question of how significant sexual dimorphism can arise within monogamy is at issue. Confronting the logical problem of how many coevolutionary "steps" are required to establish Fisherian runaway selection for such traits, she proposes a new conceptual model that simplifies the process by reducing the steps into a few general rules or "programs" for preexisting preferences. Thus, when novel features arise by mutation, they may find a highly favorable behavioral landscape in the tastes of the opposite sex. She discusses some recent experimental studies of Zebra Finch (*Poephila guttata*) preferences, where the novel phenotypic characters involved are the color of plastic leg bands (indisputably not under genetic control).

In the fourth chapter, Cynthia Bluhm provides the first experimental demonstration of the biological importance of free mate-choice in a monogamous bird (or any other bird, for that matter). Female Canvasbacks allowed to choose their own mates from a large captive population bred copiously, whereas those assigned mates not only refused to breed but vigorously rejected the males. This study has intriguing implications both for sexual selection theory (although we still must wonder *why* it makes such a difference!) and for the judicious management of captive breeding programs in other taxa.

Michael Anderson's field study clearly shows that the frequency of apparent monogamy is very high in wild Canvasbacks (99.1%), but not without its hidden complexities. A few birds (7%) switch mates between first and second (last) broods of the season, which might qualify them as "serial monogamists" (or "serial polygamists," depending on your semantic preference). Anderson also identifies the ways in which males contribute modest postzygotic PI and typically forego the mixed strategies characterizing so many other waterfowl. Finally, he discusses a variety of ecological factors that seem to constrain male reproductive alternatives and may have led to the evolution of this version of monogamy.

The next two chapters offer extensive use of the comparative approach on which ethology was founded, but in exactly opposite ways to get at the same subject. First, Frank McKinney extends the variations-on-monogamy theme by intensively reviewing a single duck genus, *Anas*. The "dabbling ducks" are relatively well studied as a group, exhibiting the full spectrum of monogamy from single-season bonds with considerable extra-pair (including forced) copulations to essentially permanent bonds with high mate-fidelity. By contrasting northern and southern hemisphere species, McKinney shows the ecological correlates of variable male investment and how reproduction options depend on the availability of alternative mates.

Bruce Beehler's approach differs in that he studied the exception, rather than the rule. Birds of paradise, which provide classical examples of secondary sexual characters derived from polygyny, also include some little-known monogamous species. Beehler compared one of these, the Trumpet Manucode (*Manucodia keraudrenii*), with two of its sympatric polygynous relatives in a New Guinean forest. He found that the three differ in ways strongly implicating diet as a major constraint on male Manucode reproductive options. With considerable data on the availability of the preferred foods, he argues that the Manucode's specialization on particular fruits may have forced biparental care and thus monogamy. These data on the relationship between diet and male investment are a most welcome contribution to the monogamy literature, in the John Crook (1964) behavioral ecology tradition.

Finally, Bertram Murray contributes a short chapter that raises the neglected issue of demography's influence on mating systems. He uses the Lotka population model to consider, from the "floater" female's perspective, the choice of entering a polygynous relationship vs that of deferring until the next season and breeding monogamously. Through an extreme (and therefore illustrative) example, he shows how expected longevity must affect such decisions and points out that populational phenomena (including mating systems) are the collective result of presumably adaptive individual choices. A much-expanded treatment of how life history patterns shape mating systems will appear elsewhere.

CHAPTER 2

MULTIPLE PARENTAGE AND APPARENT MONOGAMY IN BIRDS

PATRICIA ADAIR GOWATY[1]
Department of Zoology, University of Oklahoma, Norman, OK 73019

ABSTRACT.—Use of electrophoretic exclusion techniques for descriptions of kinship between care-giving adults and putative offspring in apparently monogamous birds had led to observations of multiple paternity and maternity (more than one father or mother represented in a brood). The exclusions suggest questions about the social mechanisms leading to multiple parentage, most of which probably will not be resolved by further descriptions based on electrophoresis. Uncertainty of maternity is possibly an overlooked but important driving force in the evolution of behavior of female birds. Concepts of mating systems based on genetically effective matings (those resulting in offspring and, therefore, evolutionarily significant) are stressed as alternatives to concepts employing primarily adult breeding-season dispersions, "cohabitations," pair-bonds, or ecological factors. The concepts derived from gametic contribution ratios explicitly focus on genetically effective mating unlike alternatives that tacitly assume high positive correlation between mating and "cohabitation." The critical evolutionary question is how the complex overt aspects of social organization affect and are correlated with the genetically effective mating pattern and vice versa. Tasks for ornithologists include estimates of population size, sex-ratio, and reproductive success of potentially breeding adults before current sexual selection theories for the evolution of mating systems can be evaluated critically.

INTRODUCTION

To evaluate theories for the evolution of mating systems including monogamy, both individual reproductive success and kinship must be known (e.g., Hamilton 1964; Orians 1969; Trivers 1972). Ornithological studies typically rely on circumstantial evidence such as association patterns in order to evaluate these variables. Thus recent studies of mating systems' evolution are based on long-term field studies of individually marked animals, which allow observation and precise description of association patterns and assumed genetic relatedness. This paper addresses some issues associated with individual reproductive success and kinship in "monogamous" species that arose from an attempt to describe empirically the mating system of Eastern Bluebirds (*Sialia sialis*) in genetic terms.

Here I first define the topics, multiple parentage and apparent monogamy, and describe an example of an apparently monogamous species that exhibits multiple parentage of clutches and broods. Second, interpretations of multiple parentage in relation to apparent monogamy are examined, that is, what are the possible social pathways leading to multiple parentage in apparently monogamous birds? This section includes predictions of how behavior should vary under the alternative pathways to multiple parentage. The final section of the paper briefly examines concepts of mating systems that stress gametic success and raises a fundamental question, how are apparent mating patterns as represented by social

[1] Present address: Department of Biological Sciences, Clemson University, Clemson, SC 29631.

dispersion related to patterns of gametic success? Empirical knowledge about the origins of multiple parentage and the social pathway leading to it in relation to social organization is limited. Therefore, I hope readers will focus on the legitimacy of the questions and reserve conclusions until appropriate data are forthcoming.

Apparent Monogamy and Multiple Parentage

Apparent monogamy.—Apparent monogamy is an overt aspect of social organization; it is what one observes readily, viz. the dispersion of adults during breeding attempts (operationally, one male-one female social units), which might be called patterns of "cohabitation" or association. Apparent monogamy indicates little about biparental care or mating exclusively, two frequently assumed correlates of monogamous social organization. So defined, apparent monogamy is a simple operational descriptive term that probably includes the vast majority of birds (Lack 1968). So defined, it is the least common denominator of the many avian species called "monogamous" (Lack 1968).

For example, although apparent mating patterns such as monogamy are usually considered to reflect actual mating patterns, theory (Trivers 1972) and data (e.g., Erickson and Zenone 1976; Evans 1982; Gowaty and Karlin 1984) suggest that apparent mating patterns and actual (genetically effective) mating patterns may not be the same, and may in fact be two independent variables of social organization. "Apparent monogamy" indicates that some of the assumptions usually made in relation to "monogamous" birds perhaps should be reformulated explicitly as questions about them.

The concept of apparent monogamy does not challenge the widely accepted notion of monogamy as a prolonged association and essentially exclusive mating relationship between one male and one female (Kleiman 1977; Wittenberger and Tilson 1980). Wittenberger and Tilson (1980) imply by the quoted definition that occasional covert matings outside the pair-bond do not negate the existence of monogamy. However, their notion of monogamy is fully operational for only a small number of bird species largely because research in avian mating systems has emphasized polygyny and polyandry (Oring 1982). Most studies have been conducted on unmarked populations and have been too short to substantiate either exclusivity in mating or prolonged association (even within one breeding season). The term "apparent monogamy" (operationally, one male-one female social units) emphasizes questions about the relationship between mating or gametic success and dispersion or association patterns in social organizations.

It should also be kept in mind that concepts of mating systems (Lack 1968; Emlen and Oring 1977; Ralls 1977; Gowaty 1981a; Daly and Wilson 1983) are statistical and populational even though mating is an act between individuals. These facts have led to confusion over whether individuals or populations are polygynous, polyandrous, or monogamous. These difficulties can be resolved by recognizing that mating systems are epiphenomena: ways of understanding the combined outcomes of individual mating events (see Murray, Chap. 8, for additional clarification). Mating systems are not things in and of themselves; they are not emergent properties (properties unpredictable from observation of components of that which is being studied: Salt 1979) of populations, but are thought constructs that possibly facilitate a better understanding of evolution within populations. So, in some definitional schemes (see below), the combined outcomes of many individual mating choices lead to a collective characterization (statistical or categorical summarizations of the behavior of the components of that which is being studied: Salt 1979) of mating for a population. In keeping with these ideas, Wickler and Seibt (1983) stress that the preferred use of the term "monogamy" is a mating tactic of an individual, and that monogamy as a sociographic unit and monogamy as an individual's mating tactic should be distinguished. My use of "apparent monogamy" is an attempt to make an additional distinction between what is known and assumed about gametic success.

Multiple parentage.—Multiple parentage occurs when a clutch or brood has more than one mother or more than one father. Multiple parentage has been observed in some birds (Vehrencamp 1977; Bertram 1979; Andersson 1983; Koenig et al. 1984) and inferred in others (Evans 1982; Alatalo et al. 1984; Gowaty and Karlin 1984). Multiple maternity can be verified when more than one female is known to lay in a single nest (e.g., Vehrencamp 1977; Koenig et al. 1984). Multiple paternity can be inferred when more than one male copulates with one female (e.g., Beecher and Beecher 1979). Alternatively, electrophoresis of blood proteins offers a more direct inference of multiple parentage than observations of copulations (Sherman 1981 and references therein). Multiple parentage is important in the context of questions about gametic contributions because such observations afford important evidence that extra-pair copulations, for example, may be evolutionarily effective.

In this paper, particular stress is given to multiple parentage, because techniques (such as electrophoresis of blood proteins for "paternity" exclusions) designed to yield direct evidence of gametic success lead to observations of broods or litters with more than one father and/or mother. Such data seldom allow assignment of biological parentage. Thus the social pathways leading to multiple parentage are open to investigation and these alternate hypotheses for multiple parentage are stressed below.

Multiple Parentage in Apparently Monogamous Eastern Bluebirds

Eastern Bluebirds consort in pairs (Gowaty 1980, 1981b, 1983); however responses to experimental manipulations of both behavior and habitat are inconsistent with existing theory on the evolution of monogamy. For example, experimentally deserted (lone), female Eastern Bluebirds are as reproductively successful as control (paired) females, implying that males may have time for extra-pair copulations (Gowaty 1983). When more than one nest site is available within a defendable territory, a male can attract more than one female to a territory, implying that apparently polygynous nesting attempts might occur when nesting is confined to natural cavities. In an effort to describe gametic contributions of care-taking adults (putative parents), I collaborated with an ecological geneticist who examined electrophoretically detectable variability at two loci in 257 birds from 57 broods representing 40 families of Eastern Bluebirds (families often consist of two or more consecutive broods) breeding in northwestern South Carolina in 1981 (Gowaty and Karlin 1984). We originally called the study a paternity exclusion study (reflecting our general expectations); it was also a maternity exclusion study. One of 20 (5%) sampled males took care of at least one offspring not his own and four of 27 (15%) sampled females took care of at least one offspring not their own. Multiple parentage occurred in 25% of the subsample of families for which complete data were available (i.e., families in which the electrophoretic phenotypes, assumed genotypes, for the care-giving male and female and all nestlings were known). These estimates must be considered conservative, i.e., estimates of minimal frequencies of multiple parentage, for a variety of important reasons including that the technique can only exclude parentage (Gowaty and Karlin 1984). The conservative nature of exclusion tests must be emphasized; it is likely that we were unable to detect many more actual cases of multiple parentage. Thus we concluded with the working hypothesis that the genetically effective mating pattern of Eastern Bluebirds may be described best as polygamous because we do not yet know the social pathways that resulted in multiple parentage. Existing theories for the evolution of monogamy now seem inadequate because explanations of the female tactic of intraspecific nest parasitism are largely unavailable.

Explanations for Multiple Parentage

Asymmetries in relatedness between care-giving adults (putative parents) and resident offspring depend on the social pathways leading to multiple parentage (Table 1). Predictions about how putative parents should behave if these social mechanisms have evolutionary effects are discussed below in relation to the possible kinship asymmetries.

Multiple paternity.—Multiple paternity may be caused by multiple matings by resident females, forced copulation, multiple-bond matings, or egg-dumping.

Extra-pair copulation by females.—Extra-pair copulation (EPC) by females may lead to (1) no offspring, (2) multiple paternity of her own brood, or (3) multiple maternity of someone else's brood (a mixed strategy of offspring care). When a female engages in EPCs, she may mate with two or more males while cooperating in parental care with only one male (she may mate polyandrously). If all copulations are effective (eggs are fertilized), she may produce a clutch fathered by more than one male, multiple paternity, in which case the care-giving male would be unrelated to at least some of the offspring. Alternatively, she may deposit eggs fertilized by these extra-territorial males in nests outside the territory of her pair-bonded mate. In such cases, care-giving males can be related to all of the offspring for which they care even though the female will care for only some of the offspring she produces—a mixed strategy of offspring care by females (see later section). Strong selection pressure against males that cooperate with females in caring for clutches and broods sired by more than one male is probable (Trivers 1972; Barash 1976; Morton et al. 1978; Zenone et al. 1979; Power and Doner 1980; Power et al. 1981; Gowaty 1981b). EPC by females implies that the mating strategy of some females (at least) is polyandrous by choice.

Forced extra-pair copulations.—Forced extra-pair copulation (FEPC) is well described in ducks (see reviews in McKinney et al. 1984 and 1983 for references). If multiple paternity results from FEPCs, the resident males will not be related to the young resulting from such matings. Selective pressure against males that care for offspring resulting from FEPCs of their mates should be strong and should favor males that guard their mates from forced copulations (Barash 1976; Beecher and Beecher 1979; Power and Doner 1980; Power et al. 1981; Gowaty 1981b).

TABLE 1

MECHANISMS OF MULTIPLE PARENTAGE AND ASYMMETRY OF RELATEDNESS OF CARE-GIVING ADULTS TO OFFSPRING. "+" INDICATES GENETIC RELATEDNESS; "−" INDICATES NON-LINEAL RELATEDNESS

Explanations for multiple parentage	Relationship of care-givers to offspring	
	Male	Female
Multiple paternity:		
Forced copulations "FEPC"	−	+
EPCs by females	−	+
MBMs by females	−	+
Multiple paternity and maternity:		
Intraspecific egg-dumping	−	−
Multiple maternity:		
EPCs by males	+	−

Multiple-bond matings.—Multiple-bond matings (MBM) can also lead to multiple paternity. Consider the following scenario. If male Spotted Sandpipers (*Actitis macularia*) return to find their old mates laying with a new mate, there is more than an even chance that the original male will displace the new one (Lewis Oring, pers. comm.). The female is, in fact, paired to two different males at different points in her clutch generation and this obviously could lead to multiple paternity of her brood. Neither the basic sociographic unit nor the individual mating tactic of females may differ under this social pathway. Nevertheless, the genetic consequence is like that for polyandrously (by choice) mating females.

Egg-dumping.—Egg-dumping is difficult to observe but known in some passerines (Bullough 1942; Seel 1968; Yom-Tov et al. 1974; Weatherhead and Robertson 1978) and well-described in ducks (see McKinney, Chap. 6, for references). Females laying their eggs in the nest of other females may be unpaired "floaters," females that have lost their nests, or paired females resident on other territories (Yom-Tov 1980). Egg-dumping by paired females has attributes of a mixed strategy of offspring care (discussed below). In this paper, the term egg-dumping will refer exclusively to cases of multiple parentage in which neither the resident female nor resident male are lineal relatives of young resulting from a dumped egg (Table 1).

Although egg-dumping may not seem parallel to the three cases above of EPC, FEPC and MBM, each of which leads to multiple fathers of (presumably) one female's clutch, all four social pathways can lead to multiple paternity. Thus, unless eliminated by appropriate kinship and behavioral data, each should be considered a viable alternative hypothesis for multiple paternity.

Although egg-dumping is infrequently considered in relation to mating tactics of individuals, it clearly could be. For example, a female may copulate with multiple males, produce a singly sired clutch and dump eggs from EPCs, thus "stealing" this misdirected parental care from other females and males rather than from her bonded mate. Such occurrences could have important implications for our understanding of individual reproductive tactics.

Multiple maternity.—Multiple maternity occurs when females lay eggs in nests not their own. Depending on whether these females have copulated with the resident males, multiple maternity can be due to egg-dumping (as defined above) or to EPCs by resident males.

Extra-pair copulations by males.—Usually discussions of EPCs by males assume that the extra-pair females lay any resulting eggs in some other males' nests. It is possible that extra-pair females lay their eggs in the nests of the males with which they copulate. In either case, the females laying eggs in nests not their own will be gaining parental care from (probably) unrelated females. If multiple maternity is from EPCs by resident males, the care-giving male would be related to the progeny from such copulations and the care-giving female would not (Table 1). EPC by males implies that the mating tactic of some males, at least, is polygynous.

Trivers (1972) predicted multiple mating by males in which a "monogamous" male invests in the offspring of a primary female, without passing up opportunities to inseminate other females which he will not aid. Trivers' discussion focused on the behavior of males mating with females that lay in the nests of other males, tactics that also might be known as mixed strategies of offspring care. Similarly, I think it possible that (a) a male may mate with two or more females, (b) both of whom may lay eggs in his nest, but (c) the male shares the parental care effort with only one of these females, thus misdirecting her parental care.

Although seldom discussed, such uncertainty of maternity may occur in a variety of avian species. Selection pressure against exploited females is expected if only by analogy to the well-discussed theoretical expectation for selection against males that care for offspring not theirs (e.g., Trivers 1972). If stealing parental care from females occurs regularly, countermeasures should have evolved also and the following predictions should hold: (1) female-female aggression will be greatest during the egg-laying period of the nesting cycle (Gowaty 1981b); (2) females will guard their nest-sites, especially during egg-laying; (3) nest-site guarding will vary in intensity depending on the number of females residing in a territory (an analogy to the polyandrous and monogamous Acorn Woodpecker [*Melanerpes formicivorus*] groups studied by Mumme et al. 1983b); (4) nest-site guarding by females will be stronger than by males; and (5) conflict between males and females over the intensity of nest-site guarding will depend on their independent probabilities of kinship to offspring.

Despite the fact that the numerical advantage associated with a male's mating with two or more females, each of which lays eggs in the same nest, may be lacking in the above example, probable advantages to multiple maternity of clutches for males include: (1) increased variability of progeny; (2) fertility assurance; and (3) increased genetic quality of offspring.

Categories of Misdirected Parental Care and Uncertainty of Maternity

No generally accepted word deals with stealing of parental care from females (Table 2), reflecting the assumption that it is difficult to manipulate females into caring for unrelated offspring in their own nests (see Power 1984 and Gowaty 1984 for alternate opinions about the semantic issues). True as the assumption may be for mammals, it is probably relatively easy to manipulate a female bird into caring for unrelated offspring in her own nest (witness the success of interspecific egg-dumping as an obligate or facultative parenting strategy).

TABLE 2

CATEGORIES OF CARE OF NON-KIN: THE EVIDENCE, THE EVOLUTIONARY LOSER (THE VICTIM) AND THE COMMON NAME OF THE MECHANISM OF CARE OF NON-KIN

Evidence	Victim	Name
Multiple paternity	care-giving males	"cuckoldry"
Multiple maternity	care-giving females	—
	males and females	egg-dumping (brood parasitism)
Multiple maternity and paternity		a) intraspecific
Eggs or chicks of another species		b) interspecific

In a recent study of colonially nesting Cliff Swallows (*Hirundo pyrrhonota*) up to 24% of the nests contained the eggs of more than one mother (Brown 1984). The careful observations lead to the conclusion that the eggs belonging to parasites may require less incubation time than host eggs and to the hypothesis that parasites may toss eggs from the host nest. Brown's paper stresses that intraspecific nest parasitism may be a cost peculiar to colonial nesting. Such vulnerability may follow from the egg-laying habit itself and is probably widespread even in dispersed nesting passerine species. Unlike mammals, birds probably lack maternity certainty because of egg-laying. This lack of certainty of maternity for birds has been overlooked as a potential driving force of female behavior (see the predictions about female behavior in the preceding section).

Modes of evolutionarily misdirecting (stealing) care of offspring have important similarities. In Table 2 egg-dumping is designated as interspecific or intraspecific because interspecific brood parasitism is functionally identical to intraspecific egg-dumping (Hamilton and Orians 1965; Payne 1977). When egg-dumping (as specifically differentiated in this paper from EPC by females and males) occurs, both the resident male and resident female care for unrelated offspring, and the selection pressures against egg-dumping should be similar (Hamilton and Orians 1965; Payne 1977).

CONCEPTS OF MATING SYSTEMS

What are the implications for mating system theory if social pathways to multiple parentage indicate non-monogamous mating tactics by individuals? That is, how many matings outside of one male-one female social units must be effective (resulting in offspring) for individuals to be polygynous (males) or polyandrous (females)? Clearly, only one in each case is necessary. Whether such variation is evolutionarily interesting depends upon how many "covert" matings outside the one male-one female social unit must be effective for a (so-called) monogamous mating system to be polygynous or polyandrous? The answer to that question is unclear. If 5% or more of males have more than one female on their territories, some workers designate a species "polygynous" (Verner and Willson 1969; Carey and Nolan 1979). A similar arbitrary criterion, depending on the frequencies of extra-pair copulations, may serve to label apparently monogamous species as effectively polyandrous or polygynous.

Observations of multiple parentage arising via social pathways that indicate variation in mating tactics in apparently monogamous birds can confound concepts and definitions of mating systems, because there are several, often implicit, ideas in the question, "What is the mating system?" Wickler and Seibt (1983) provide an excellent and timely review of such sources of ambiguity. The traditional question is concerned with consort patterns or pair-bonding patterns (e.g., who is with whom on which territory and what are the behavioral mechanisms that facilitate close association?). Studies of mating systems usually are based on the assumption that there is high correspondence between consort patterns (or pair-bonding patterns) and patterns of effective matings, the gametic contributions. When multiple parentage is observed, the second question, whose crux concerns genetically effective copulations, becomes paramount. Attempts to characterize mating systems in terms of evolutionarily effective results have been made in relation to sexual selection theory (e.g., Ralls 1977; Wade and Arnold 1980; Gowaty 1981a; Daly and Wilson 1983) and recently these characterizations of mating systems have been favored by Wickler and Seibt (1983). However, few empirical attempts to describe avian mating systems in terms of gametic success have been attempted (Gowaty and Karlin 1984).

During David Lack's time definitions of mating systems referred to the quality and duration of pair-bonds (Lack 1968). Such questions remain interesting, if different, from other questions about mating systems. Others (e.g., Emlen and Oring 1977) have emphasized ecological factors that effect dispersions of breeding adults; thus terms like "resource defense polygyny" gained sway. Ecological definitions seem to orient thinking about some aspects of social organization more productively than those based on the duration or quality of pair-bonds (partly because ecology is described and measured more precisely than pair-bonds). However, neither pair-bonds nor dispersions of breeding individuals may be as highly correlated with mating patterns as usually thought. For example, Red-winged Blackbird (*Agelaius phoeniceus*) females bonded to vasectomized male red-wings frequently may mate with more than one male (Bray et al. 1975). Thus labelling red-wings as polygynous may be misleading because the label tends to obscure the fact that some females, at least, may mate with more than one male. Although the dispersion of breeding adults appears polygynous, the genetically effective mating pattern and the genetically effective result may differ. How are dispersion patterns or pair-bond patterns correlated with patterns of gametic contributions by females and males? Any observed differences may lead to important alternate conclusions about the evolution of sexually selected traits. Indeed, the only data capable of rejecting or unambiguously confirming sexual selection hypotheses are those indicating individual gametic success.

Gametic contributions.—Concepts.—The following sections highlight mating systems' concepts based on genetically effective results (i.e., matings resulting in progeny) rather than those based on consort patterns, pair-bonds, or strictly ecological factors. Such definitions (see below) are "temporary verbalizations of concepts" (Mayr 1982) that reflect modern, conceptual reorientation toward such topics as sperm competition, extra-pair copulations, and notions about the effects of actual rather than putative kinship (Sherman 1981). The concepts explored here emphasize the possible discrepancies between genetically effective mating and other correlates of social organization.

"Genetic polygyny" may be defined as occurring when one individual male effectively contributes gametes to the progeny of more than one individual female (an individual's tactic) so that the ratio of effective matings by males to effective matings by females is <1 (Ralls 1977; Gowaty 1981a). Similarly, "genetic polyandry" occurs when one individual female effectively contributes gametes to the progeny of more than one individual male so that the ratio of effective matings by individual males to effective matings by individual females is >1. And, "genetic monogamy" occurs when the ratio = 1. These ratios are gametic contribution ratios (CGRs) (Gowaty 1981a) and can be used to describe both individual tactics and collective characterizations of populations (mating systems). For example, within a population one mating subgroup (one male and one female) may have a GCR = 1, whereas another in the same population (for example, one male and three females) has a GCR < 1. So despite the fact that all of the females and some of the males mate with only one individual (i.e., are monogamous) the name that best describes the statistical phenomenon (combined outcomes of matings) is genetic polygyny or GCR < 1.

Definitions of mating systems (collective characterizations of individual mating tactics) based on gametic contributions are simple. Only three categories are known: =1, <1, and >1 ratios. Degrees or gradations of either <1 or >1 would be apparent; for example, mating systems predominating in polygynous individuals (GCR < 1) can be relatively strong or weak depending on the magnitude of the ratio of breeding females to breeding males. When the GCR = ½, on the average one male mates for every two mating females in the population; when the GCR = ¼, on the average one male mates for every four mating females and so on. Thus, for example, for GCRs < 1, the presumed evolutionary advantage to mating males is inversely proportional to the magnitude of the ratio (see Daly and Wilson 1983 for an alternative viewpoint).

Definitions of mating systems based on GCRs depend on a measure of effective mating success between individual mating males and individual mating females. The salient comparison is between reproductive success (RS) of individual mating males and the RS of individual mating females, an approach that stresses intersexual competition.

This conceptual scheme appears not to address systems such as "polygyny-polyandry" occurring in various ratites (Jenni 1974; Bruning 1974). In rheas and tinamous apparently individual males mate with more than one female and individual females mate with more than one male. But the combined outcomes of individual mating success for females and males are unknown and may lead overall to any one of three possible GCRs (1, <1, >1). It remains that the GCRs may be quite different from the patterns of dispersion of breeding adults and the pair-bonding. In addition, if the GCR equals one, it would lead to the interesting conclusion that the genetic result of sociographic polygyny-polyandry may have some important similarities to monogamy. Also in the case of sociographic promiscuity, where there is an equal probability that every individual could mate with every other individual (indiscriminate mating), the GCR could also equal one, again leading to the conclusion that sociographic promiscuity and monogamy may have some important genetic similarities. GCRs alone cannot be used to estimate the intensity of sexual selection (Wade and Arnold 1980), but combined with accurate estimates of the adult sex-ratios, these intensities can be estimated. On the other hand, GCRs will unambiguously indicate whether the genetic systems are advantageous to females or males. Orians (1969) indicated that the evolution of mating systems is dependent on the individuals gaining the least advantage in a mating system; thus the concept of male vs female advantage is an important one, which GCRs emphasize.

Definitions of mating systems based on differences in the variance in RS of males and of females have also been suggested (Daly and Wilson 1983) and are based on the theoretical prediction that in apparently polygynous species the variance in male RS is greater than the variance in female RS (Bateman 1948; Trivers 1972). In schemes using definitions based on the ratios of variance in male RS to variance in female RS, non-mating individuals are emphasized (i.e., variance in male RS may be large because some males do not mate at all). The theoretical notion that variance differences in RS between the sexes drives the evolution of many complex behavioral and morphological traits is so cogent (see Ralls 1976, 1977 for alternative conclusions) that it is easy to forget that such differences have seldom been verified (Bateman 1948; Payne 1979; Howard 1979; Clutton-Brock et al. 1982) or evaluated in terms of traits resulting from mating advantage (Howard 1979). Thus it is appropriate to predict, rather than conclude, for example, that one male-multi-female social units lead to greater variance in male RS than female RS and that such variance differences may result in traits such as sexual dimorphism.

More tangible examples may illustrate why it should be useful to evaluate variance differences directly rather than on the basis of assumed RS. Red-winged Blackbirds are labeled as polygynous because many nesting attempts are apparently polygynous (although others are apparently monogamous) (Verner and Willson 1969; Orians 1980). If the sex-ratio of potentially breeding Red-winged Blackbird adults is 1:1 (and all females mate effectively), it follows that the variance in male RS will be greater than variance in female RS. However, if the sex-ratio of potentially breeding adults favors females, as the sex-ratio of Red-winged Blackbird nestlings does (Fiala 1981), variance in male RS may be neither large nor different from variance in female RS. On the other hand, in apparently monogamous species, variance differences in RS between the sexes might be different if females outnumber males or males outnumber females. In the case where females outnumber males, variance in female RS might be larger than variance in male RS. These differences have not been definitively evaluated even in Red-winged Blackbirds, thus underscoring the difficulty in obtaining the appropriate data. Another, perhaps more pithy example comes from a benchmark study of Bobolinks (*Dolichonyx oryzivorous*) by T. A. Gavin and E. K. Bollinger (pers. comm.). Based on detailed behavioral observations and electrophoretic exclusion data, they report an intriguing case of two males defending contiguous territories. One (B) is the only bearer of a rare allele in the adult population and has no females nesting on his territory. The other (A) has one female nesting on his territory. This female fledged six offspring, three of which carried the rare allele of male B while alternate alleles at other loci excluded male A from paternity. Two other nestlings from this brood could not have been fathered by male B and could have been fathered by male A. The sixth nestling could have been fathered by either male. Using traditional measures of RS, male A would have been assigned a RS value of 6 whereas male B would have been assigned a RS value of 0. Based on the exclusion data, male A had a RS value of 2, possibly 3, whereas male B had a RS value of 3, possibly 4. This situation came from a species that is sociographically polygynous!

To evaluate sexual selection hypotheses for the evolution of social organizations, reliable estimates of population size (including "floaters"), the sex-ratio, and RS (including estimates of genetically effective extra-pair and multi-bond copulations) of potentially breeding adults are all needed. None of these measures is easily obtained, yet knowledge of all of them is crucial to evolutionary analysis of social organizations.

The major advantage of gametic approaches to thinking about mating systems is that they focus attention on evolutionary mechanisms. With these definitions it is possible to ask how genetically effective mating patterns affect social organization patterns such as consort patterns. Mating is correlated with rather than synonymous with the dispersion of breeding adults and pair-bonding behavior. Gametic success, bonding, and social dispersion should be thought of as independent variables. Secondary questions can also be posed: How do copulatory patterns (including null copulations, those not producing progeny) effect the dispersion of adults during the breeding season? These definitions help set priorities. For ornithologists two tasks seem paramount: (1) reliable estimation of gametic success, and (2) estimation of numbers of territorial non-breeders and floaters.

ACKNOWLEDGMENTS

I thank A. A. Karlin for our collaboration and early discussions of what our data might mean. D. Heckel, D. Mock, J. Quinn, R. Tilson, and B. Vestal read and commented on previous drafts of the paper. I thank especially R. Tilson for insightful comments and discussions about these topics throughout 1983.

CHAPTER 3

THE ORGANIZATION OF BEHAVIOR AND THE EVOLUTION OF SEXUALLY SELECTED TRAITS

NANCY BURLEY

Department of Ecology, Ethology and Evolution, University of Illinois, 515 Morrill Hall, 505 S. Goodwin Ave., Urbana, IL 61801

ABSTRACT.—Two sorts of programs, or evolved plans for making decisions, are envisioned to control sexually selected behaviors. One of these, the general assessment program (GAP), is hypothesized to interpret information and determine the best course of action for its bearer. GAPs are organized by rules of strategy that accord with the principles of sexual selection. Pattern recognition programs (PRPs) are species-specific programs that identify and classify signals and cues, including novel ones. Because of their species-specificity and because of the arbitrariness of many signals, the configuration of PRPs is less predictable than that of GAPs. By viewing behaviors as organized in this manner, it becomes possible to appreciate how a sexually selected trait might evolve rapidly, given a minimum number of mutations.

To determine whether organisms do have programs favoring the rapid evolution of sexually selected traits, animals must be exposed to novel stimuli and their reactions recorded. Several relevant experiments are reported here. In the first set of experiments, adult Zebra Finches (*Poephila guttata*) were allowed to perch next to their choice of four opposite-sexed conspecifics, three of which were attired in novel phenotypes. Colored leg bands that covered a portion of the orange legs provided the alteration in phenotype. In one experiment, females were given their choice of bandless, vs red-, orange-, and green-banded males. They preferred to perch next to red-banded males and were unattracted to green-banded males. In another experiment involving silver, yellow, and black-and-white bands, females were attracted to males attired in yellow. Red is a color normally limited to the beaks of male Zebra Finches, whereas yellow is not present on the body surface of wild-type Zebra Finches. The preference of females for yellow-banded males underscores the relative unpredictability of preferences for novel traits. Subsequent experimentation revealed that females were probably attracted to the contrast provided by yellow bands on orange legs and not to the yellowness *per se*. In similar experiments, males were attracted to black-banded females and were not attracted to blue-banded ones.

In the second category of experiments, birds were permitted to display perching preferences for same-sexed birds for which heterosexual experiments revealed a significant preference. Females preferred to perch next to blue-banded females and were unattracted to black-banded ones. Similarly, males preferred to perch next to green-banded males and were unattracted to red-banded ones. It appears that red-banded males and black-banded females are simultaneously perceived as attractive by opposite-sexed individuals and possibly as threatening by same-sexed individuals. If these novel traits had a genetic basis, the highly organized, preexisting responses of conspecifics would enhance the possibility that their frequency would increase and that further evolution of the species appearance could occur without necessitating any mutation that affects the behavior of individuals of either sex.

INTRODUCTION

This paper examines the existence and possible significance of organized responses to novel sociosexual stimuli in a monogamous estrildid, the Zebra Finch (*Poephila guttata*). A line of reasoning is developed to suggest that these responses may permit some understanding of the process of sexual selection in monogamy, and in particular the origin of "aesthetic" (non-functional) species characteristics. Unfortunately, the importance of sexual selection in monogamous populations is not well understood. Indeed, although sexual selection has long been thought to operate in polygynous and promiscuous species (e.g., Campbell 1972; Ghiselin 1974; Halliday 1978; Blum and Blum 1979; Thornhill and Alcock 1983), its occurrence in monogamous species is often ignored or denied (but see O'Donald 1973, 1974, 1977a, b). This viewpoint that sexual selection is unimportant in monogamy typically results from consideration of the potential for great variation in reproductive success of males of polygynous species, where alleles strongly favored by sexual selection can increase rapidly in frequency (e.g., Arnold 1983 and references therein). It is worth recalling, however, that sexual selection, like natural selection, operates on differential reproductive success, and that if sex differences and evolutionary change occur in monogamous populations, which they clearly do, then sexual selection is likely to occur as well. I focus here not on how variable reproductive success is effected in monogamy, but rather on origins of aesthetic traits that might enhance the quality (if not quantity) of mating opportunities of bearers of such traits. Monogamous birds vary from being highly dimorphic to sexually monomorphic or indistinguishable (Burley 1981a) in appearance. They also display great variability in their degree of "flashiness," with only some species conforming to the dull image that might be expected on the basis of their mating habit. Numerous historical and ecological factors have doubtlessly contributed to this diversity. It is my hope, however, that this paper will stimulate readers to reflect on the diversity in species phenotype displayed by monogamous birds and to ponder the role that sexual selection has played in effecting that diversity.

In this paper I argue that much of our understanding of behavioral evolution is based unnecessarily on rather simplistic conceptions of gene function and somewhat naive expectations regarding ordered sequences of mutations. My purposes here are: (1) to suggest that understanding the evolution of behavior requires viewing units of behavior as complicated entities ("programs") that permit the expression of emergent properties, by which I mean evolved plans of procedure for making decisions and implementing behaviors; (2) to hypothesize that such programs operate to allow individuals to react appropriately to environmental stimuli depending on their sex, age, and situation; and (3) to hypothesize that when novel social stimuli occur (through mutation), individuals of both sexes are already programmed to respond in organized ways. Programs are assumed to have complex genetic bases. Their physiological components are not considered here, although some components will be obvious.

I focus here on a subset of social behaviors, namely sexually selected behaviors and morphological traits associated with them. I hypothesize the existence of two types of programs (General Assessment Program and Pattern Recognition Program) to regulate such behaviors. Supporting evidence is provided for the hypothesis that latent sex-specific reactions to novel phenotypes occur in organisms, and that these latent responses form preadaptations favoring the increase in frequency of some possibly small set of mutations (and disfavor others) when they occur. Hence the evolution of sexually-selected traits may be very rapid, not requiring an ordered sequence of mutations to effect novel, complex behaviors.

The Problems

Much of the recent literature on the evolution of behavior is subject to at least two kinds of criticism. (1) Many theorists, in particular mathematical modelers, treat complex behaviors as if they were determined by single genes (e.g., Levitt 1975; Macnair and Parker 1978; Parker and Macnair 1978; Wade 1979; O'Donald 1977a, 1980; Lumsden and Wilson 1981). However, single genes do not determine morphological properties such as chins and elbows (Waddington 1957), and numerous genes probably affect seemingly "simple" properties such as body size (Lande 1978, 1980b, 1981b; Falconer 1981; but see Parsons 1980). Although the ultimate origin of evolutionary novelty resides in mutation, it is reasonable to doubt conclusions based on overly simplistic conceptions of gene function. If behavioral traits such as mate preferences and altruistic tendencies have complex genetic bases, how meaningful are predictions based on single-locus models with alternative alleles for selfishness and altruism or selectivity versus non-selectivity?

A closely related issue concerns the context-specificity of gene function. Within species, optimal reproductive strategies are often, if not universally, different for the two sexes. However, it is unlikely that differing strategies are programmed by sex-specific genes because, regardless of the mechanism of sex determination, conspecifics of both sexes share most loci in common (Ohno 1979). Moreover, although sex-limited expression of traits sometimes occurs, research on the process of sexual differentiation and gender role acquisition (e.g., Harris 1970; Money and Ehrhardt 1972; Quadagno et al. 1977; Bancroft 1978; Gorski 1979) has revealed the existence of considerable developmental flexibility for numerous traits. Also, many factors that might affect reproductive strategy, such as physical condition and age, vary over the course of an individual's lifetime. These considerations further erode the plausibility of the idea that individual genes effect complex behaviors.

(2) Theorists often construct models suggesting that an ordered sequence of mutations is necessary to arrive at an evolved condition. If an ordered sequence is required, then the possibility of the sequence occurring declines rapidly as the number of steps in the sequence increases (Frazzetta 1975).

A simple scenario for the evolutionary convergence of intrasexual and intersexual signals demonstrates both of these problems. Assume that a mutation occurs to give male bearers an advantage in intrasexual competition. The presence of this mutant allele can be detected by conspecifics. Non-carriers of the allele that respond by avoiding interactions with conspecifics bearing the mutation benefit more than non-carriers that "ignore" the competitive asymmetry, because the former incur less risk of serious injury. However, because a benefit to avoiding carriers could not accrue prior to the spread of the mutation conferring competitive superiority, an "avoidance" mutation can increase in frequency only after the mutation conferring competitive advantage does so. (The "avoidance" mutation may be recurrent, but will only be selected for after the competitively superior genotype becomes reasonably abundant.) Similarly, an allele for female preference for the competitively superior phenotype could only be selected for after the superior male type has become established in the population. Once females begin to prefer males based on the appearance of superiority rather than an actual test of superiority (male combat), carriers of the competitively superior allele will gain an additional benefit, viz. they will not have to prove their superiority as often. By this reasoning, two traits that, if present, would enhance the possibility of fixation of the allele for competitive superiority (avoidance of the phenotype by other males and preference for the phenotype by females) are not likely to occur until after bearers of the competitive phenotype have become well established. Of course, in taxa in which appropriate responses to novel social conditions can be easily learned, mutations for an avoidance response and/or preference response may be unnecessary.

Eventually the carrier phenotype may become fixed in the population, making obsolete the allele for preference of the phenotype. Later, a new mutation might arise that further enhances competitive ability in another way. Females might show an initial aversion to carriers of the new mutation, having evolved a preference for what is now the status quo. If so, the mutation sequence must now begin anew, with the added necessity of a mutation to suppress the previous female preference or a modifier mutation that changes the expression of the previous mutation.

If single genes function in the highly specific fashion suggested by the above scenario, the potential for sustained evolution of sexually selected signals appears quite restricted. However, we know that homeostatic mechanisms of behavior, physiology, and development exist permitting organism adjustment to differing ecological circumstances or physiological milieux (Thompson 1942; Goss 1965) and suggesting the possibility of a much more flexibile response. The immune system, for example, appears designed to respond to the unexpected: "A given antibody is a theory made by the animal about what is in its environment" (Steele 1981). I argue below that similar complex mechanisms exist that facilitate organized responses to novel sociosexual environments.

To replace the above scenario, I suggest that many genes interact to run programs that contain instructions of the sort: "If you are a dominant male, do thus and so, whereas if you are a young male, do this instead, and if you are a female in reproductive condition, pursue a third alternative." Programs are coded for by genes common to the gene pool. Individuals then make context-specific responses. Such programs eliminate the need for several of the repeated steps in the sexual selection scenario outlined above; instead, they permit individuals to respond to novel stimuli in the context of their sex, stage in life, and fighting ability. These programs also allow us to understand how evolution through sexual selection could proceed rapidly without requiring that most mutations have large effects on behavior. They also provide a simple mechanism for the convergence of intrasexual and intersexual signals (Darwin 1871; Lande 1980a, 1981a).

This paper focuses on programs related to sexual selection and contains an hypothesis for the existence of two kinds of programs with discrete characteristics. One program (the General Assessment Program—see below) contains an organized set of rules that form a strategy for optimizing mating and reproductive opportunities. These rules generally accord with sexual selection theory (insofar as our understanding of the process of sexual selection is accurate), but are refined to reflect ecological and life history characteristics of the population/species concerned. The function of the other program, Pattern Recognition Program, is to identify, classify, and evaluate sociosexual stimuli. Its role is principally classificatory rather than strategic. As a result, its structure is not derived directly from evolutionary theory and its rules cannot be easily predicted from such theory.

Sexual Selection and the Evolution of Sexual Dimorphism

Darwin (1871) hypothesized that sexual selection involves two interrelated processes: within-sex competition for mates and between-sex selection of mates. Darwin believed that females were typically the "selective sex," and males the "competitive sex," but current theory (Trivers 1972) holds that the form and intensity of sexual selection varies with the relative apportionment of parental investment (PI) between the sexes. Where females incur most or all the costs of rearing offspring, access to females is the factor most limiting male reproductive success (RS). In this instance, female-female competition for mates and male choice should be largely absent. Where PI is more evenly shared by the sexes, both sexes should experience similar levels of competition and display similar patterns of mate selectivity (e.g., Burley 1977a, 1981b).

Mate preferences can be broadly categorized into several types. *Economic* traits confer direct, resource-related benefits to the mate and/or offspring of the individual possessing the characteristic. For example, size is an economic trait if it reflects differential ability to obtain and maintain a breeding territory (Orians 1969; Searcy 1979a, b). *Genetic* traits are those that indicate high genetic quality of the selected individual, with the likelihood that the genetic quality of offspring will be relatively high. Vigor and resistance to disease are examples of traits that could be indicative of high genetic quality. A current tendency to discount the utility of mate choice for genetic traits (Williams 1975; Maynard Smith 1978; Borgia 1979) is based partly on the expectation that the heritability of fitness is too low for females to profit from discrimination of genetic traits. However, recent evidence suggests that the heritability of fitness components may often be substantial (e.g., Istock 1978; Giesel and Zettler 1980; Grant and Price 1981; Boag and Grant 1981; Cade 1984).

Economic and genetic traits both have proximate functional value; i.e., by being selective for such traits, individuals improve the quantity and/or quality of offspring. By contrast, aesthetic traits lack proximate value. Aesthetic traits are arbitrary symbols whose value is defined by the signalling system of a particular species. Among the most common aesthetic traits are those associated with species and sex identification. In one species of bird the message, "I am an adult male," is communicated via bright red plumage, whereas in another it is conveyed by iridescent blue feathers. Of course, such traits are ultimately functional if they increase the reproductive success of their bearers, but the benefit to the bearers does not explain the origin of the preference for the trait. The origin of aesthetic mate preferences is currently a topic of considerable interest (Lande 1981a; Arnold 1983 and references therein).

Fisher (1930) provided an hypothesis for the evolution of aesthetic traits. His hypothesis (hereafter referred to as "Fisherian selection") posits that aesthetic traits evolve from economic traits. For example, large males can defend good territories, thereby increasing their mates' reproductive success. A mutation conferring female preference for large males is therefore advantageous to its female bearers. An important assumption is that preference is directional: the largest males are always preferred. (Recent quantitative modeling by Lande [1981a] suggests that this assumption can be relaxed somewhat.) This results in a runaway process favoring ever-increasing male size. At some point large size becomes dysfunctional with respect to its original purpose, but the preference for it remains through the momentum of female choice: a female that selects a non-extreme mate produces sons that are unattractive to almost all females and so, while surviving, fail to reproduce. Size is eventually limited by decreased survivorship of the largest males.

The Fisherian scheme can easily be conceived to originate with genetic as well as economic traits. Assuming that there is a heritable basis to traits that indicate health, longevity, vigor and stamina, females should prefer those males judged to be most physically fit in order to maximize offspring quality. The Fisherian process would just exaggerate such traits past the point at which they are reliable indicators of physical capacity. However, Fisherian selection, important as it is, cannot explain all occurrences of aesthetic features. The hypothesis (and recent modifications and alternatives by Zahavi 1975, 1977; Borgia 1979) may have only limited relevance to species that form long-term pair bonds, because monogamy limits, but does not eliminate, variation in reproductive success among males, the driving force of Fisherian selection (Mayr 1972). Nevertheless, it is clear that aesthetic features, as defined above, are present in many monogamous species, although they do not appear to be as exaggerated as in many non-monogamous species. Moreover, it is often difficult to imagine how flamboyant or "bizarre" traits that appear to have only display value evolved from a "functional" (economic or genetic) state (Borgia 1979).

This reasoning suggests that aesthetic traits often become evolutionarily elaborated with only minimally functional antecedents. One likely way in which this may happen is through the amplification of signals involved in species recognition and reproductive isolation (Sibley 1957; Mayr 1972). In this process, co-occurring species evolve phenotypes that emphasize differences; individuals possess mechanisms that enable them to recognize the opposite sex but prefer conspecifics with traits that exaggerate the species' identity. To see one possible origin for such preferences (see also Muller 1942; Kaneshiro 1980), we can consider the model of allopatric speciation. (This model is used for simplicity only; the reasoning that follows is not closely tied to mode of speciation.) Two populations have been separated for a sufficiently long time to have diverged genetically. Upon reestablishing contact both show post-mating isolation, although the phenotypes and behavior of the two groups are similar. Small phenotypic changes have nevertheless occurred, perhaps as a result of genetic drift. Suppose, for example, that during isolation the body surface of individuals of population A obtained a bluish cast, whereas individuals of population B became green. When contact is reestablished, individuals of population A that display a preference for blue mates and those of B that display a preference for green mates will have a reproductive advantage over individuals that mate randomly. This preference should operate at a comparative level; i.e., individuals that are the most green or most blue of those available should be preferred, because their population of origin is least in doubt. Hence, over evolutionary time individuals of population A will become bluer and those of B greener. It is not at all clear that directional preferences would be lost once the "species" diverge sufficiently to become unambiguous; in fact, given that individuals of one or both sexes tend to prefer the most extreme mates, individuals with less extreme preferences may suffer reproductively because their offspring would be relatively unattractive. Thus, what begins as a functional (genetic) preference may quickly acquire aesthetic value.

The reproductive isolation hypothesis might be criticized on the grounds that evidence for morphological (usually size-related) character displacement is weak (Grant 1972, 1975). However, the evidence for reproductive character displacement, as opposed to ecological character displacement, is strong (Grant 1975). Researchers have demonstrated that color and pattern are important aspects of species recognition in birds (e.g., Lack 1943; Klint 1980), and Smith (1966) has shown that sympatric congeners are sensitive to minor but discrete species differences. Character displacement is an important component in our understanding of speciation processes (Carson 1968; Ohta 1978; Kaneshiro 1976; Powell 1978; Templeton 1979). The mate choice component of sexual selection is likely the force that typically leads to reproductive character displacement.

Fisherian selection and the reproductive isolation hypothesis provide some understanding of the possible origins of aesthetic features and the evolution of sexual dimorphism. Other processes, such as selection for rapid pair formation among nomadic or temperate species (O'Donald 1972; Jehl 1970; Burley 1981a), may also be involved. If traits evolve directly through aesthetic preferences (i.e., without genetic or economic precursors), it seems probable that preferences for particular traits must often be present prior to the occurrence of the trait in the population. Otherwise, when a mutation occurs, it is unlikely to establish itself except through drift or pleiotropy. In the following sections, I address the questions: (1) how sexually selected traits arise that have primarily aesthetic significance; and (2) how sexually selected traits can evolve without requiring a series of ordered mutations for morphological and behavioral traits.

The General Assessment Program (GAP).— The purpose of the General Assessment Program (GAP) is to permit organisms to respond strategically to situations involving sexual competition and mating/reproductive opportunities. The GAP receives information (from another program; see below) and decides what, if anything, should be done in response to that information given the context of the individual in whose body the program "resides." For example, suppose an unmated adult female encounters a group of bachelor males. Her GAP will respond by examining a mate assessment subprogram to evaluate the relative qualities of the various males. If an adult male encounters the same group, his GAP will turn on a different subprogram, one regulating intrasexual competition.

If something like the GAP does exist, a general pool of information must be available to all individuals in a population. The implications for the evolution of sexually selected traits are straightforward: the occurrence of GAPs provides preadaptations favoring certain mutations. For example, a mutation that confers even a slightly threatening male phenotype might confer a double advantage to its male bearers: a survival advantage (because other males avoid bearers of the mutation) and a reproductive advantage (because females are attracted to such bearers). It is unnecessary to invoke the occurrence of a separate mutation that allows females to act as if they perceive that mutant males are threatening to other males, because females have the same genetic background and programs for behavior as males. The program simply tells females to respond to the same cue differently than males.

Two questions immediately follow from this conceptualization: (1) Where did GAPs originate and how do they evolve? Sexual selection is an ancient process which has operated on organisms at least since the evolution of anisogamy. Indeed, it is likely that the battle of the "sexes" was responsible for the very origin of anisogamy (Parker et al. 1972). The rules governing sexual competition probably began to be incorporated into the genomes of sexually reproducing species shortly after the evolutionary invention of sexuality. Over evolutionary time, these rules have become amplified, refined, and adapted to a wide variety of species "needs." Neither microevolutionary events, speciation, nor larger macroevolutionary events eliminate GAPs. Instead, following changes in prevailing selection pressures, GAPs are modified. Different species have different GAPs, but their general form is predictable from sexual selection theory. (2) Under what circumstances will novel ("mutant") phenotypes be perceived as "attractive" and/or "threatening" to certain conspecifics? To answer this question, I conceive of another kind of program with very different characteristics.

The Pattern Recognition Program (PRP).—The Pattern Recognition Program (PRP) responds differentially to environmental stimuli; it identifies and classifies important stimuli from the multitude of those available and it makes a preliminary assessment of the meaning of novel stimuli. This determination depends on the sensory system of the organism and how its past history has shaped the species PRP. PRPs are further organized by physiological processes such as stimulus filtering and result in such phenomena as releasers and supernormal stimuli (e.g., Tinbergen 1942; Lorenz 1950 and references therein). The evolution of signals of all kinds necessitates changes in and/or refinements of PRPs.

In my conceptualization, PRPs permit the identification of possible mates and competitors. GAPs then use this information to decide what to do given the context (sex, reproductive state). Whereas GAPs are relatively consistent from species to species, PRPs tend to be more species-specific. For example, consider the task of species identification. Imagine two closely related sympatric species with similar mating systems and reproductive biology. Both species use visual cues, primarily color, to locate potential mates. Allow premating isolating mechanisms to be well developed, with species A typically having a green appearance while species B is usually blue. It follows that the reaction of the PRP in species A to blue individuals will be negative (because blue individuals tend not to be appropriate mates), whereas the reaction to green individuals will be generally positive. The PRP of species B will effect the opposite reaction. On the other hand, the GAPs for mate choice of the two species could be identical; i.e., the species could have very similar strategies of reproduction.

PRPs are not only species-specific. The form and evolution of PRPs involved in sociosexual interactions (and possibly other behaviors as well) are relatively unpredictable, both because signals (in the sense of Burghardt 1970; Otte 1974) are somewhat arbitrary symbols and because of the lack of predictability of neurophysiological response. For example, imagine that degree of redness signals dominance rank in a particular species of fish. Orange individuals are relatively subordinate, whereas those that are cherry red are dominant. We can then ask the question: if a mutation occurred that conferred a slightly purple tinge to the red coloration, would this be perceived by the fish as an intensification of red? To predict the response of the fish to the mutation we would at least need information on the specificity of the red receptors to different wavelengths and on the occurrence (or lack thereof) of blue or purple receptors. If such receptors were present we would need to know the social significance of blue when present on the body surface of conspecifics. In short, without a great deal of specific information, no reliable prediction of a mutant's social status would be possible.

At this point, it is possible to appreciate how PRPs permit the evolution of novel signals as well as the amplification of existing ones. Consider the Red-winged Blackbird, the adult male phenotype being glossy black with red epaulets on the wings. Males also have a narrow yellow stripe at the distal border between the red epaulet and the black body-and-wing plumage. Red epaulets function in successful territory defense in this species (Smith 1972) and are very conspicuous during display. Imagine that the ancestor of this species had a similar appearance, but that the degree of redness varied more and the yellow border was absent. Male-male competition to acquire and defend territories involved displays of the epaulets, and those with the most intense red had the greatest competitive advantage. Then imagine that a mutation occurred that made a part of the epaulet yellow. Such a mutation could make the epaulet look less intensely red (more orange), particularly if yellowness were spread in small dots throughout the red patch ("yellow dot" mutation). However, if the yellowness occurred at the border between red and black ("yellow border" mutation), its effect could be to make the epaulet stand out against the black background or appear more intensely red. A yellow dot mutation would probably be selected against, whereas a yellow border mutation might be selected for, because it could promote the ability of its bearers to intimidate other males. However, in this case, it would not be the yellowness *per se* that is important; rather its evolution as a discretely observable characteristic was a byproduct of directional selection for more intense redness. Other mutations (perhaps conferring a "green border" or a "white border," as found in the Tricolored Blackbird, *Agelaius tricolor,* Peterson 1961) might have had the same effect.

We can extend this line of reasoning even further, back to a time when blackbirds were merely black. At that time, males displayed ownership of their territories by erecting their iridescent black feathers, assuming a certain posture and singing a particular song. The males that appeared largest, shiniest, and most active were most convincing to other males and were most successful at acquiring and maintaining territories. Could we predict that a mutation that conferred a patch of red color in a relatively conspicuous place would be perceived as intimidating to other male blackbirds? I am confident we could not. However, neither could we be certain that such a reponse would not occur. The possibility of a response would depend on the birds' ability to perceive the color red, of course, but also on more subtle characteristics that determine the birds' response to the color red in that particular context.

Allowing for the possibility that blackbird color receptors, sensory filters, and neural wiring were arranged so that a displaying red-shouldered blackbird was more intimidating than a plain blackbird, there are, however, certain predictions we could make: having received this information, individual blackbirds would respond to it differently. The most dominant males should be least intimidated by the pre-mutant phenotype advertising greatest dominance and hence least intimidated by a mutant. In contrast, the least dominant males should be most intimidated by a mutant. The dominance status of any mutant might be enhanced by the mutation, but the status of an otherwise relatively dominant individual would increase more than the status of an otherwise subordinate because other dominants would be more likely to be intimidated by the display of a dominant-cum-epaulet than of a subordinate-cum-epaulet (Shields 1977; Ketterson 1979; Rohwer and Ewald 1981). If individual females made mate choices in part on the basis of the dominance displays of males, they would also be affected by males with the normal phenotype, but their GAPs would inform them that the appropriate response would be attraction to mutant males, just as it had previously told them to be attracted to the most dominant pre-mutant phenotype. In other words, GAPs would incorporate this species-idiosyncratic response to redness into the normal domain of decision making. Further mutations might occur that result in modification of the location or intensity of the coloration, or even add other new colors to the body surface, depending upon the responsiveness of PRPs.

In summary, General Assessment Programs (GAPs) for mating behaviors evolve in accordance with the principles of sexual selection. Through the occurrence of GAPs, organisms of both sexes have organized ways of responding to novel features of their environment, and the length of the sequence of ordered mutations necessary to evolve any sexually selected trait is curtailed. Pattern Recognition Programs (PRPs) are more idiosyncratic among species, in part because sensory systems are sensitive to only a portion of available stimuli and emphasize some stimuli while disregarding others. The sensory system of any species will therefore be preadapted to perceive some not-yet-evolved stimulus in a particular way. Through the conjunction of PRPs and GAPs, species sustain directional selection for sexually selected characteristics which are largely or totally of an aesthetic nature.

Evidence for GAPs and PRPs

There is a growing literature on both empirical and theoretical aspects of sexual selection. Rather than exhaustively review that literature here, I will provide two examples of apparent decision-making rules of the kind expected to be components of GAPs.

I (Burley 1977a, b) extended Trivers' parental investment hypothesis to predict patterns of intraspecific variability in mate selectivity, arguing that when both sexes incur PI, individuals will exercise selectivity in proportion to their own mate quality: highly preferred individuals should be more selective of mates because they can afford to be so, whereas less desirable individuals must settle for inferior mates or fail to reproduce at all. When there is a discrepancy between the sexes in relative PI incurred, the sex with greater PI should show higher overall selectivity and less variability in selectivity than the sex investing less PI. In species in which only one sex incurs PI, a correlation between mate quality and selectivity should be absent: because the sex with no investment should mate indiscriminantly, the reproductive opportunities of lower quality individuals of the investing sex are not constrained by their attractiveness. This hypothesis should be modified to include mating investment (Low 1978; Alexander and Borgia 1979; Willson and Burley 1983), because if the act of securing a mate restricts an individual's ability to locate future mates, it should be somewhat selective of mates regardless of a lack of PI (e.g., Johnson 1982).

To test hypotheses regarding between-sex and within-sex variability in selectivity in a mating system with biparental care, I introduced Rock Doves (*Columba livia*) to an experimental design in which an unmated individual was presented with a choice between two opposite-sex birds tethered to eliminate intrasexual interference competition (Burley 1977a,1981b; Burley and Moran 1979). Because female pigeons have greater PI than males, I predicted that, overall, females would show greater selectivity and that there would be less intrasexual variability in selectivity among females. Both hypotheses were supported. Pigeons behave as if guided by a general, three-step decision-making rule: "Monitor one's own attractiveness to opposite-sex conspecifics and set one's selectivity on the basis of attractiveness and sex."

Another example concerns the hangingfly (*Bittacus apicalis*) investigated by Thornhill (1976). In this and related species, males offer food items to females during courtship. Females partially or totally devour these "nuptial gifts" during copulation, which varies in length from less than one to more than 30 min. Thornhill found that duration of copulation was positively correlated with size (length × width) of nuptial prey over a portion of the prey size range (3–19 mm^2), but for large prey (20–55 mm^2) no such correlation was found. Duration of copulation also affected the number of sperm transferred. Below about 5 min, no sperm were transferred; between 5 and 20 min there was a strong positive correlation with number transferred; beyond 20 min, however, the number of sperm transferred did not increase. Both sexes could terminate copulation, at which time the male was likely to grab the food from the female, possibly to use it again to court a new female. Thornhill found that copulations involving prey less than 18 mm^2 were always female-terminated, whereas those involving larger prey were male-terminated. Males bearing small or unpalatable prey items were often totally rejected by females, but when only small prey items were available, females often did accept males but copulated for only short intervals.

Apparently a GAP for evaluating the quality of potential nuptial gifts is common to both sexes of adult hangingflies. When male hangingflies capture prey items smaller than 19 mm^2, they usually feed on them and then discard them. They retain only larger items for courtship. Sometimes this rule is broken and smaller prey items are used in courtship; this tendency would likely be affected by the availability of larger prey items and a male's success in capturing large prey. Females react similarly, showing a preference for males with large prey and rejecting or displaying partial acceptance of males with smaller prey. Again, it seems reasonable to expect that the acceptability of males with smaller prey is dependent on the availability of males with larger items.

Circumstantial evidence for GAPs can be obtained whenever organisms are found to follow consistently certain patterns of behavior consistent with theory. Because PRPs are conceived to be unpredictable, evidence for their occurrence cannot be obtained so easily. Moreover, evidence that different organisms perceive the world differently will not tell us much about how differential perception affects evolution through sexual selection. *To test the hypothesis that species are preadapted to respond in certain ways to novel phenotypes we must present them with novel phenotypes and measure their reaction.* To have much chance in identifying an attractive novel phenotype, experimenters either need a large number of available mutations (probably prohibitive for most species) or they must rely on "pseudomutations" by experimentally altering the organism's phenotype (e.g., Noble 1936; Noble and Vogt 1935; Marler 1955; Lewis 1971; Smith 1972; Rohwer 1977). Below I summarize evidence that individuals of one species, the Zebra Finch (*Poephila guttata*), find certain oppositely-sexed pseudomutants "attractive" (compared to wild-type) and others "unattractive." More importantly, I demonstrate that reaction to pseudomutants is sex-specific, suggesting that a program (GAP) apparently exists permitting individuals to respond to novel conspecific phenotypes in the context of their sex.

Zebra Finches are monogamous Australian estrildids. They are sexually dimorphic, the female having a gray dorsal surface, an off-white ventral surface and an orange beak. Males' beaks tend to be redder and males also have black breast bars and narrow, horizontal black-and-white striping extending from the chin to the breast bar, gold cheek patches, and chestnut spotted flanks. Both sexes have species-specific, black-and-white markings on the face and tail (see Keast 1958 and Immelmann 1965 for more thorough descriptions). The experimental pseudomutations consisted of several colors of small plastic leg bands, of the kind routinely used for identification of individual birds in the field and laboratory. Preferences for pseudomutants versus wild-type (unbanded) birds were determined by relative time spent perching next to each of four types: an unbanded bird, and three birds each banded on both legs with one band of a particular color. Both legs per bird were banded with the same color, but each banded bird wore a different color. Experiments were conducted using an apparatus that permitted viewing by experimental birds of only one of the four stimulus birds at any one time. Birds could engage in courtship activities and could communicate by sound as well as sight. Between trials, color bands were rotated among stimulus birds so that observed preferences for color bands could not have resulted from preferences for particular birds. (See Burley et al. 1982 for details of experimental design and additional detail on experiments 1 and 2.)

Experiment 1.—Individual females were tested for their tendency to perch next to unbanded (+) males or those wearing red (R), orange (O), or light green (g) bands. Perching preferences were significantly non-random (Friedman multisample test, $P < 0.001$); females most preferred R males, least preferred g males, and did not discriminate between O and + (nonparametric multiple comparisons tests [Fig. 1]).

Experiment 2.—Individual males were tested for their tendency to perch next to unbanded (+) females or those wearing black (Bl), orange (O), or light blue (b) bands. Perching preferences were significantly non-random (Friedman multisample test, $P < 0.001$); males spent most time perching next to Bl females, least time with b females. They did not discriminate between O and + [nonparametric multiple comparisons tests (Fig. 1)].

Limited evidence suggests that these preferences have a sexual component,

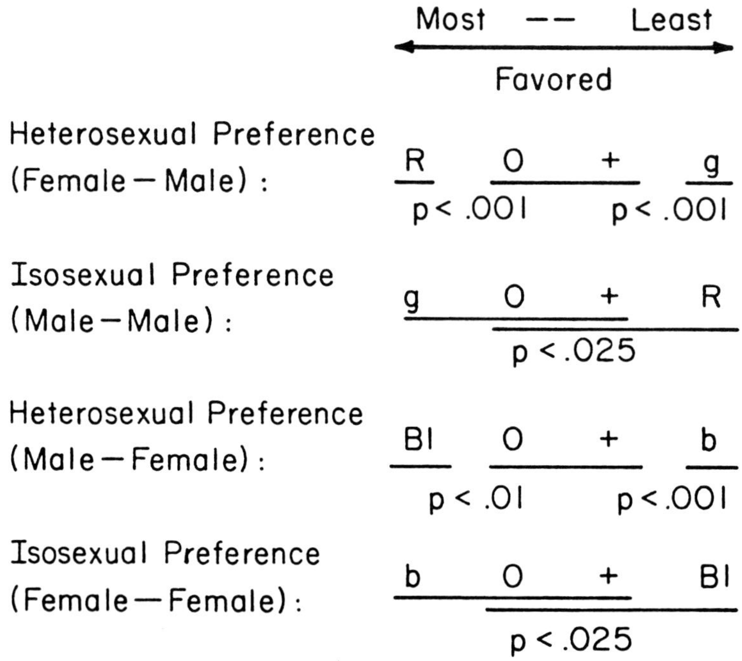

FIG. 1. Summary of heterosexual and isosexual perching perferences. R = red-banded; O = orange-banded; g = green-banded; + = unbanded; Bl = black-banded; b = blue-banded stimulus birds. Breaks in the horizontal bars underlining the stimulus sequences denote statistically significant differences in perching preferences (a posteriori tests).

because test birds were deprived of physical and visual contact with opposite-sex conspecifics for some time prior to the experiments and they commonly engaged in courtship activities during experiments. Also, number of courtship song bouts sung by experimental males correlated positively with time spent in view of stimulus females in the one experiment in which data were collected on song behavior (Burley et al. 1982).

At least two alternative explanations might account for these patterns: (1) the responses are a measure of general (context-free) color preferences. A priori, it seems unlikely that the birds would have consistent color preferences that are truly context-free. If this were the case, we would find, for example, that female Zebra Finches would prefer to sit in red chambers versus green ones and prefer red food over green food. Perching preferences of birds for empty, color-coded chambers have not been measured; however, Zebra Finches apparently prefer seeds dyed with light green food color over those dyed red (unpub. data). (2) The preferences are purely social. If this were the case, male Zebra Finches would prefer to perch next to R vs g males and females to prefer to perch next to Bl vs b females. Data presented below indicate that this is not the case.

Allowing the inference that the experimental procedure reflects sexual (i.e., mate) preferences, it remains to be asked what kind of information the pseudomutants "signal" that makes them differentially attractive or unattractive. One possibility is that the preferences are aesthetic, that rules exist common to the majority of individuals of one sex that attract them to opposite-sex individuals, but that novel phenotypes are not perceived to amplify existing signals. Another possibility is that the pseudomutants are perceived to advertise exaggeration(s) of existing signals. If so, what could those signals be? Possible signal meanings include state of health, species identification, and/or relative dominance status. It seems reasonable to implicate species identification in these results because estrildid leg color is interspecifically variable and Zebra Finches inhabit a wide geographical range that overlaps with those of a number of related species (Immelmann 1965; Burley et al. 1982). Moreover, the leg and beak color of their closest relative, *P. bichenovii*, is blue-gray with a metallic cast. Light blue is a color not preferred by male or female Zebra Finches when worn by the opposite sex (above and Burley et al. 1982).

State of health and dominance status are likely to be interrelated. A trait that may advertise either or both of these conditions is beak color. Beak color varies from orange yellow to brilliant red; males' beaks tend to be redder than females', but there is considerable overlap between the sexes (Burley and Coopersmith, pers. obs.). Beak color also varies with age: dependent offspring have black beaks. Young birds in adult plumage (3–4 months) often have paler beaks than those of older birds (6–7 months or more). Among fully adult birds housed in unisexual cages, individual beak color sometimes changes measurably (using a Munsell® color set) within the span of one or two weeks. Among color-banded birds permitted to pair and reproduce in a large aviary, substantial color change occurs over one to three months. Changes in value, chroma, and hue have been detected and occur in both directions (whiter/blacker, paler/brighter, more yellow/more red). Thus, beak color is not a constant trait, but rather varies over time and may reflect changing social and/or physiological conditions. In males, the size of the black breast bar also varies, although somewhat more slowly. The cause and significance of variation in the size of the breast bar are unknown.

The color of the red leg bands employed in experiment 1 is similar to a very bright red beak of a male Zebra Finch. Red bands on legs may intensify the effect of the beak, possibly also signalling better health or higher dominance status. Perhaps the presence of light green bands diminishes the effect of the red beak. It is more difficult to discern a link between the female phenotype and males' preference for black-banded females. As noted, black is the color of the juveniles' beak, but males do not seem to prefer females with black beaks (Garson et al. 1980). Black is found in both the species-specific horizontal tail stripes and vertical eye stripes.

If the effect of the experimental pseudomutations is to alter the perceived dominance status of their bearers, and if GAPs occur, we would expect, for example, that males allowed to perch next to an array of males (R, O, g, +) would react differently than females allowed the same choice. It seems somewhat reasonable to expect males to show an aversion to R-banded birds; however, as noted earlier, this effect should be greater on subordinate individuals than on dominants. Very dominant individuals should be the most disposed to attempt to test the honesty of the signal displayed by R males. If so, dominant individuals might even be attracted to R males; therefore, we might expect variability in male response to the R, O, g, + test sequence. Similar reasoning would apply to in-

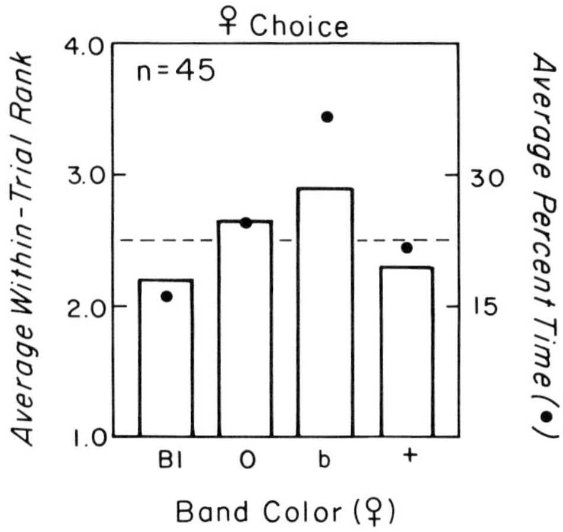

FIG. 2. Female perching preferences in isosexual experiment with black (Bl), orange (O), blue (b), and unbanded (+) test sequence.

trasexual interactions among females. If, on the other hand, pseudomutations primarily or exclusively signal state of health or species identification, I would predict either general avoidance responses toward b and g individuals, preference for association with most probable conspecifics (Bl, R), or possibly random association if, for example, in nature Zebra Finches form mixed-species flocks that include species which the b- and g-banded birds slightly resemble (Immelmann 1965).

In summary, either isosexual (i.e., same-sex) preferences in the opposite direction of heterosexual preferences, or individually variable isosexual preferences (reflecting dominance status of test birds) would support the interpretation that pseudomutations affect the perceived dominance status of their bearers. Either of these results would also tend to support the occurrence of GAPs and PRPs, in that both sexes display organized, sex-specific responses to phenotypes that presumably have yet to occur in the species. Random isosexual responses could result if the pseudomutations serve only an epigamic function or if Zebra Finches make relatively little species discrimination when joining groups for foraging or other non-mating activities. Isosexual responses in the same direction as heterosexual responses would result if preferences are exclusively social, rather than sexual, or if certain phenotypes suggest a poor state of health and were possible disease carriers. Isosexual preferences in the opposite direction of heterosexual preferences, as well as individually variable and random isosexual preferences, are consistent with the interpretation that the observed heterosexual preferences (experiments 1 and 2) are sexual (i.e., mate-oriented) in nature and that such preferences are related to the occurrence of PRPs. Random isosexual responses would not support the hypothesis that both sexes possess the same GAP. However, in the following experiments, random responses could not be distinguished from individually variable preferences that reflect the dominance status of test subjects.

Experiment 3.—Adult, female Zebra Finches reared by non-color-banded, wild-type parents were permitted to perch next to Bl, O, b, or + females during two-hour test periods. Nine females were tested three times and an additional nine

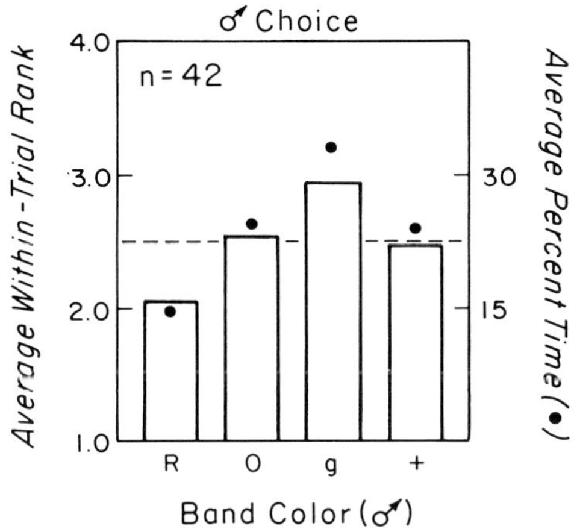

FIG. 3. Male perching preferences in isosexual experiment with red (R), orange (O), green (g) and unbanded (+) test sequence.

were tested twice. Repeated trials for each subject were spaced two or more weeks apart and involved novel stimulus sets. The tendency to perch next to the various stimulus types was nonrandom (Fig. 2; Friedman multisample test, $P < 0.05$); females spent the most time perched in view of b females (overall viewing time = 38%) and the least time with Bl females (16%). Times spent with +, and O females were intermediate. Non-parametric multiple comparisons tests revealed that the only significant difference in ranks of time spent with the stimulus types was that between Bl and b females ($P < 0.025$).

Experiment 4.—Adult male Zebra Finches were tested for their tendency to perch next to R, O, g, and + males under conditions identical to those employed in experiment 3. Eleven males were tested three times, and six males were tested twice. Perching tendencies were nonrandom, with most time spent with g males (35%) and least time with R males (15%) (Fig. 3; $P < 0.05$). A posteriori tests revealed that g was more preferred than R ($P < 0.025$), but other comparisons were not significantly different.

In both isosexual tests, results were in the opposite direction of those reported for the equivalent heterosexual test. These results demonstrate that preferences for birds wearing leg bands of these colors are indeed sex-specific. The results also indicate that both sexes are predisposed to perceive novel phenotypes in certain organized ways, thus supporting the GAP concept and the hypothesis that further evolution of the Zebra Finch phenotype through sexual selection could occur with a minimum number of mutations.

These results also have implications for understanding how color bands are perceived. Results are consistent only with the hypothesis that bands affect the perceived dominance rank of their bearers. Because test subjects had received only limited visual contact (Burley et al. 1982) and no physical contact with color-banded birds prior to testing, and stimulus birds were banded with any one color only for short periods, the responses appear to result from PRPs. The possibility also exists that multiple signal functions have converged over evolutionary time; for example, it could be that signals that perhaps originally evolved in the context

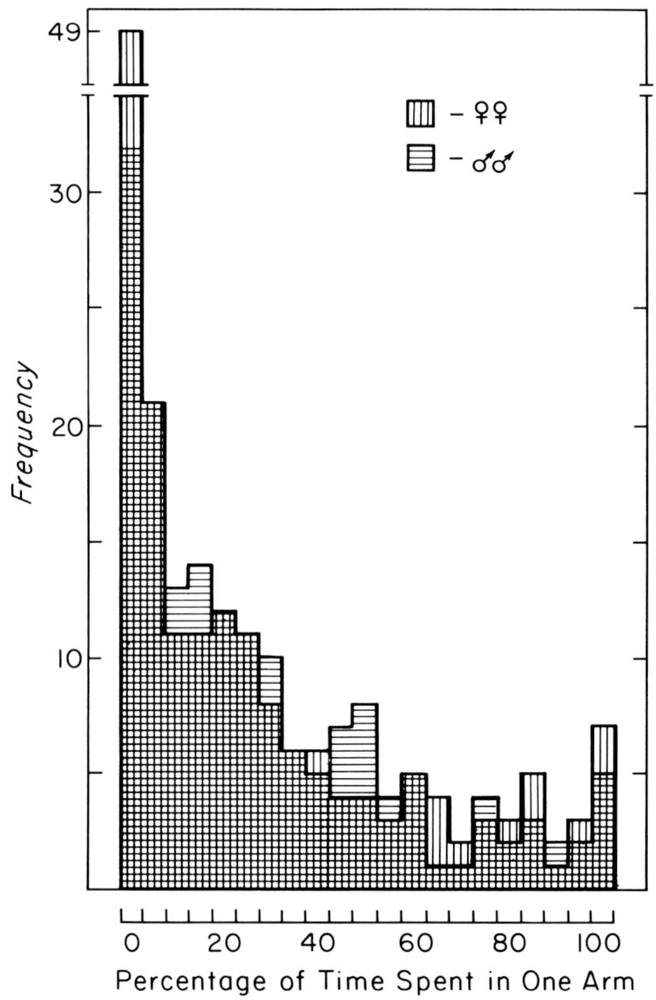

FIG. 4. Relative times spent perched in each of the four arms of the test chamber by subjects of both sexes in isosexual experiments. While perched in an arm, the subject was always able to view one stimulus.

of species recognition have been amplified for other purposes, so that an individual with exaggerated species characteristics also signals social dominance or other characteristics.

It should be noted that significance levels obtained from isosexual tests are weaker than those of the heterosexual tests. The data suggest individual variability in responses, but given the method of analysis this possibility cannot be examined without many more replicates per individual. The choice of statistical technique is limited to rank-order tests by the distributions of time spent with stimulus birds, which deviate markedly from normality (Fig. 4).

Evidence that Zebra Finches have preferences for conspecifics with novel appearances (those wearing red and black leg bands) supports the concept of PRPs. However, because red and black are colors already present on the body surface

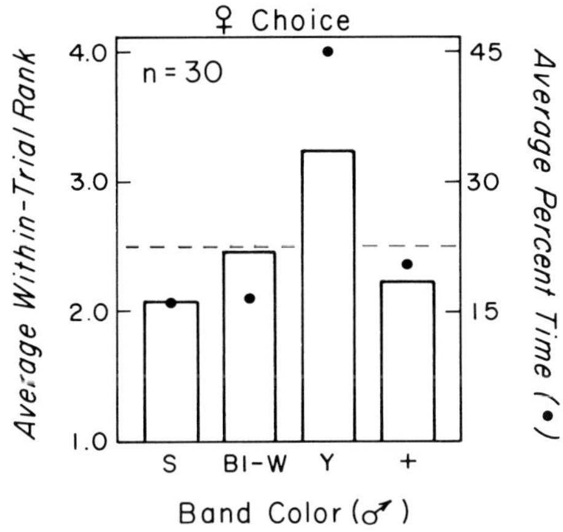

FIG. 5. Female perching preferences in heterosexual experiment involving silver (S), yellow (Y), black-white stripe (Bl-W) and bandless (+) test sequence.

of the species, the evidence shows only that existing colors could be enhanced or rearranged, but not that PRPs could function in the evolution of novel colors in the species phenotype. To determine if novel colors are generally disfavored and familiar ones preferred, I tested Zebra Finches with additional color sequences. The one I report here included dull silver (S), yellow (Y) and horizontally striped, black and white (Bl-W) bands. The dull silver resembles the predominant color of the dorsal surface of both sexes. It was selected in part to determine if birds are affected by the presence of numbered aluminum bands worn by all birds in the experiments. The narrow stripes of the Bl-W bands are similar to barring found on the "throats" of adult males. Also, both sexes have broad horizontal black-and-white stripes on their tail feathers and vertical black-and-white stripes under their eyes. The color yellow is not found anywhere on the body surface of wild-type Zebra Finches. Based partly on the results of earlier experiments, I expected females would be attracted to Bl-W males, males would be unattracted to Bl-W females, both sexes would show an aversion to Y, and that both would be relatively indifferent to S.

Experiment 5.—Adult female Zebra Finches were permitted to perch next to S, Bl-W, Y, or + males during two-hour test periods. Fifteen females were each tested twice. The methods were identical to those reported above. The tendency to perch next to the various stimulus types was nonrandom (Fig. 5; Friedman multisample test, $P < 0.005$); females spent the most time perched in view of Y males (45% of overall viewing time). They spent about equal amounts of time in view of S (17%), Bl-W (17%) and + (22%) males. Multiple comparisons tests revealed that Y was much more preferred than any other phenotype (all P's < 0.001), and that females did not discriminate among Bl-W, S, and +.

Experiment 6.—Adult male Zebra Finches were allowed to perch next to S, Bl-W, Y, or + females during test intervals of two hours. Fifteen individuals were tested twice, but two trials were discarded for technical reasons. The tendency to perch next to the stimulus types was not significantly nonrandom (Friedman multisample test, $P < 0.25$; Fig. 6). The data suggest that males prefer S and +

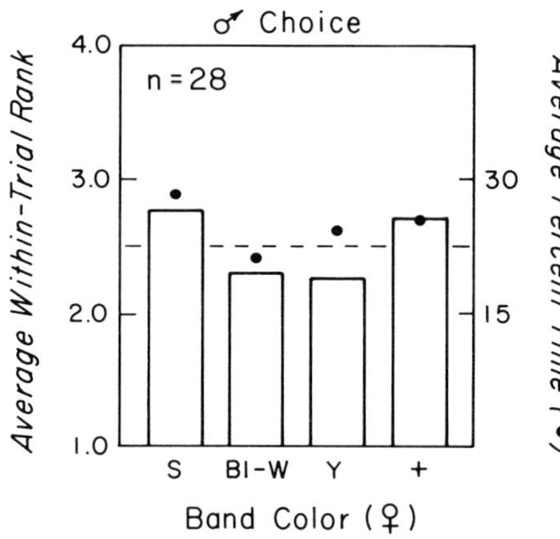

FIG. 6. Male perching preferences in heterosexual experiment involving silver (S), yellow (Y), black-white stripe (Bl-W) and bandless (+) test sequence.

females to Bl-W and Y ones, but at current levels of performance the sample size would have to be doubled to demonstrate a significant preference.

In summary, female Zebra Finches displayed a marked preference for males wearing leg bands of a color not found on the body surface of the species. This unexpected preference supports the hypothesis that totally aesthetic, "unpredictable" preferences are contained in a species PRP. It is also interesting to note that some Australian estrildids have yellow legs and/or plumage (Immelmann 1965; Bates and Busenbark 1970). Superficially, this appears to be a contradiction to the hypothesis that species identification is an important component of the preferences. However, it is also possible that, as in the Red-winged Blackbird example discussed above, it is not yet yellowness *per se* that is attractive, but rather that the contrast between orange legs and yellow bands is responsible for the enhanced attractiveness of Y males. One more experiment was performed to determine this and to test the Fisherian hypothesis that females select on a comparative basis, favoring the most extremely ornamented males available for traits known to be preferred.

Experiment 7.—Adult female Zebra Finches were allowed to perch next to males, one of which was unbanded, while the other three were ornamented in the following ways: one bird wore yellow bands (Y) as in experiment 5. Another male's toes and "legs" (tarsometatarsi) were painted with goldenrod colored Liquid Paper®, a quick-drying fluid for corrections on yellow paper. After the fluid dried, Y bands were added (one per leg). This phenotype was dubbed "super-yellow" (SY). The toes and tarsometatarsi of the final male were painted with red Magic Marker®, thus making the legs appear dark orange-red. Yellow bands were added after the marking fluid was dry to the touch. This phenotype is referred to as super-red-yellow (SRY). Males did not display any behavioral effects of the marking process, but were usually allowed to rest overnight between marking and testing. (Occasionally males had to be "touched up" an hour before a test.)

In this experiment the ornamentation of the birds in the stimulus sets could

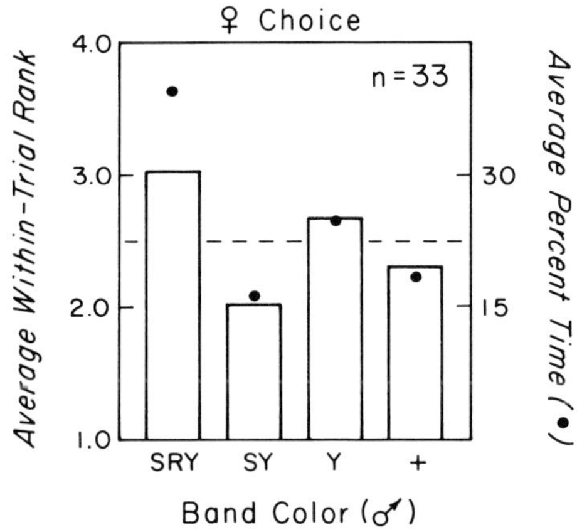

FIG. 7. Female perching preferences in heterosexual experiment involving super red-yellow (SRY), superyellow (SY), yellow (Y), and bandless (+) test sequence.

not be rotated because the marker and correcting fluid did not wear off completely enough over a short time period to make a SY or SRY into a Y or + bird. However, in two of the four perimeter sets used, the SY and SRY birds were switched after half of the trials were completed, and the Y and + phenotypes were also exchanged at this time. These two stimulus sets were used in approximately twice as many trials as were the remaining sets and were rested for several weeks between the first and second half of the trials in which they were employed. As in other experiments, each female was exposed only once to any one stimulus set.

Seventeen females were tested twice; one trial was discarded for technical reasons. Perching preferences were significantly nonrandom (Friedman multisample test, $P < 0.025$; Fig. 7). Females spent most time perched in view of SRY males (39%) and least time next to SY males (16%) and + males (18%). A posteriori tests revealed that SRY males were preferred over all other types; Y males were preferred to SY and + ($P < 0.05$); the difference in ranks of + and SY was not significant, but suggested a possible preference for + ($P \simeq 0.10$). Thus, females appear to be attracted to the contrast provided by the yellow color against orange legs and not by the yellowness *per se*. A mutation conferring yellow "ankles" (actually, the distal portion of the tarsometatarsi) would seem to have a much greater chance of increasing in frequency as a result of sexual selection than would one that conferred yellow coloration to the entire foot. This would result in the addition of spots of a novel color to the already colorful appearance of the male Zebra Finch.

Aesthetic mate preferences are usually considered to be properties of females of polygynous and/or promiscuous species, among which males provide little or no assistance to females in their reproductive attempts. Results presented here suggest that such preferences may also occur in both sexes of monogamous species with biparental care. The potential implications of these findings for our understanding of both the force of sexual selection and the dynamics of monogamous mating systems are considerable.

Discussion

Estrildid finches, particularly the Australian species, tend to be quite uniform in morphology, feeding ecology, habitat requirements, and behavior (Keast 1958; Morris 1958; Goodwin 1982). Both intra- and intergeneric hybrids have been reported for captive birds (Immelmann 1965). Hall (1962) found rather small interspecific differences in estrildid song and attributed this finding to the facts that estrildids are relatively social, typically associating in groups, and they use song primarily for courtship and do not employ it in territory defense (unlike many other passerines). Thus, he reasoned, estrildids locate mates within flocks in which species identification can be easily established through visual cues, making identification by song unnecessary.

Whether or not Hall's explanation for the conservative trend in estrildid song evolution is correct, it is clear that closely related species diverge markedly in beak and leg color as well as plumage color and pattern (Morris 1954; Immelmann 1965; Goodwin 1982). The Double Bar Finch (*Poephila bichenovii*), for example, which is probably the closest relative of the Zebra Finch (Keast 1958), has plumage characteristics that hardly resemble those of the Zebra Finch, as well as having distinctively colored beaks and legs (bluish gray instead of the Zebra Finch's red-orange; see also above). On the other hand, its courtship song is almost indistinguishable from that of the Zebra Finch (Immelmann 1965; pers. obs.). A number of the Australian species also display distinct intraspecific, regional variation in plumage color or pattern (Keast 1958). That estrildids diverge more in visual appearance than some other passerine families is indicated by the fact that plumage characteristics have long been employed as an important taxonomic tool for passerines. When Steiner (1960) used plumage differences as a primary means of categorizing 17 species of estrildids endemic to Australia, he generated 13 genera; most other classifications (e.g., Keast 1958; Mayr 1968; Goodwin 1982) recognize 5 to 7 genera.

Sibley (1957) argued that plumage traits were inappropriate taxonomic characters for certain other avian groups (passerines and non-passerines), especially highly polygynous taxa in which sexual selection is expected to be intense and those groups in which there is strong selection for species isolating mechanisms. He did not examine estrildids, but because Australian species are thought to display lifelong monogamy (Immelmann 1965), the intensity of sexual selection should not be particularly great in these species. It is also not clear what external factors might be generating strong selection for isolation. In fact, Morris (1958: 395–396) commented: "One cannot help wondering why there should be so many species of estrildines, when it seems as if a fraction of the number would have been just as efficient ecologically." Instead it seems that the radiation may be caused by factors internal to the "species" themselves.

There is growing recognition that the concepts of sexual selection and species isolation cannot be so cleanly separated as was once thought. Indeed, the potential importance of sexual selection in facilitating or causing rapid evolution and speciation is rather widely acknowledged (Carson 1968; Spieth 1974; Kaneshiro 1976, 1980; Ringo 1977; Ohta 1978; Paterson 1978; Templeton 1979; Lande 1981a, 1982). A question of fundamental importance in this regard concerns the origin of novel, aesthetic mate preferences.

I suggest that latent aesthetic preferences (contained in PRPs) become manifest when mutations occur to alter the phenotype in "attractive" ways. If this is true, then considerable evolution of species phenotype can occur without requiring mutations for changes in preferences. Isolated populations could also diverge in this way, because different mutations for phenotype would occur in separate populations. At some point in the speciation process, however, changes in PRPs must occur or premating isolating mechanisms will not evolve. Kaneshiro (1976) has suggested that changes in preferences occur because individuals in derived populations lose elements of the "mate recognition system" (Paterson 1976) through drift. Although this may occur in some cases, such an entropic process could hardly account for the vast array of phenomena which sexual selection has been hypothesized to produce.

It is clear that mutation must introduce evolutionary novelty into PRPs eventually. I envision that changes in PRPs often occur as pleiotropic effects of changes resulting from selection acting on other "programs." Consider, for example, programs that govern recognition of potential predators and suitable food items. Changes in the complexion of predators or the available food supply should result in selection on these programs. There is no reason to suppose that such selection routinely affects other programs, but some types of change are likely to have additional effects. A change in spectral sensitivity, for example, might be selected for in a species that locates predators and prey by sight. This change might also affect the PRP by altering the way conspecifics are perceived (e.g., a change that permits the perception of a color not previously perceptible to the species in question) or by altering the threshold of sensitivity to a particular color. Mutant individuals that are better able to escape predators or improve their foraging efficiency are selected for and in turn exert selection on the species phenotype. The changes in their mate preferences are not functional per se; hence they are necessarily of an aesthetic nature.

In summary, changes in PRPs may occur as the result of selection for "functional" considerations, including species isolation (see p. 27), or as the "non-functional" result of selection on other programs with pleiotropic effects on PRPs. Other possible pathways for the evolution of what I have here termed PRPs include drift (e.g., Kaneshiro 1976) and "arbitrary" mutations in PRPs (e.g., Lande 1981a; Kirkpatrick 1982; Arnold 1983).

An assumption common to several models that invoke arbitrary (non-functional) mutations in "PRPs" is that novel ("mutant") mate preferences routinely arise for which there is no counteracting preference; i.e., "non-mutant" females mate randomly with respect to a certain male characteristic whereas "mutant" females show selectivity in this respect. Considerable conservatism may be expected in mate choice. Because most novel mutations have mildly deleterious effects (Wright 1977), it would seem prudent for females generally to avoid mating with males of unusual phenotypes; I would expect such a rule to be incorporated into the GAP for mate selection. (Apparent evidence to the contrary, the phenomenon of rare male advantage is open to other interpretations [O'Donald 1977b, 1980; Burley, pers. obs.].) As a result, female preference should exert normalizing selection over a fairly broad range of traits. Male preference should be even more normalizing. If conservatism is a typical property of mate choice, males with unusual attributes that appeal to "mutant" females should not enjoy the reproductive advantages that these models indicate they should. This criticism does not obviate the possibility that some novel preferences and phenotypes evolve as suggested by such models, but it does bring into further question the generality of these modes of origin.

Finally, I believe that the conceptual compartmentalization of behaviors and, in particular, decision-making processes into "programs" with specific tasks, will prove extremely useful to the science of animal behavior (see also Gould 1982). This approach eliminates the need for simplistic notions of gene function that are widespread among students of the evolution of behavior. No paleontologist would begin a lecture on the evolution of flight by asking the audience to imagine a mutant gene for wings. Yet those interested in the evolution of behavior often ask for parallel indulgences. Perhaps our willingness to perceive behavior in such simple terms results from lack of discrete morphological structure. When one examines a wing, for example, the complex organization of this structure is readily apparent. But is the basis of a behavioral adaptation, such as altruism, so much less organized that our facile invocation of single-gene origins is more respectable? And if we invoke such genes only for the sake of argument, illustration, simplicity or metaphor, what do we really mean when we speak of them?

The need for a broader conceptualization of interacting programs and subprograms with resulting emergent properties is clearly evident if we reject the single-gene models and the necessity of the repeated occurrence of ordered sequences of mutations. The compartmentalization presented here, which deals with only a small subset of decisions organisms make and ignores the important question of how alternative possibilities are weighed, could be elaborated greatly. Although it is quite unlikely that there exists a distinct morphological "entity" that can be demonstrated to be a GAP or a PRP, the compartmentalization of the decision-making process into programs is more than a heuristic device. In this paper I have suggested one approach for testing for the existence of programs with GAP-like and PRP-like properties: ask organisms to make decisions in novel circumstances and gauge their responses. Undoubtedly other experimental approaches can be devised.

Acknowledgments

Jack Bodie, Rafe Korski, and Rich Symanski helped fashion this paper. I also thank Patty Gowaty and Doug Mock for their support, criticism, and patience. Lew Oring and an anonymous reviewer provided additional valuable comments. My research on Zebra Finches has been supported generously by the University of Illinois and also by the National Science Foundation (DEB 81-04956).

CHAPTER 4

MATE PREFERENCES AND MATING PATTERNS OF CANVASBACKS (*AYTHYA VALISINERIA*)

CYNTHIA K. BLUHM[1]
Departments of Ecology and Animal Science, University of Minnesota, Minneapolis, MN 55455

ABSTRACT.—The choice of a mate is theoretically an important determinant of an individual's fitness. The present study investigates the relationship between mate choice and reproductive success in terms of egg production in captive wild Canvasbacks (*Aythya valisineria*). The egg production of pairs allowed to form in a free-choice situation was compared with that of birds randomly assigned to each other. The prediction was that females of self-formed pairs would lay more eggs than females randomly assigned mates. The results demonstrated that *only* females of self-formed pairs laid eggs whereas females of randomly assigned pairs did not. Furthermore, the females of randomly selected pairs refused to accept their assigned mates. The large difference in egg production between the two groups indicates that free mate choice had a substantial effect on the reproductive success of individuals. Captive females recognized their self-chosen mates and acquired an exclusive and tenacious preference for them as breeding partners. The majority of the captive Canvasbacks were monogamous although a small proportion were polygynous. Pairs usually involved birds of the same age that had been reared together. Mate retention varied from 16–49% over five years and did not correlate with the previous year's nesting success.

INTRODUCTION

In sexually reproducing animals, the choice of a mate theoretically can be an important determinant of an individual's fitness (Darwin 1871; Fisher 1930). In a monogamous mating system where each breeding adult forms a pair bond with only one member of the opposite sex, mate choice may be especially important (Trivers 1972). This is because half of the genetic component and often a sizeable postzygotic parental investment for each offspring comes from the prospective mate for each breeding attempt. Mate choice is, therefore, a possible mechanism that animals might use to increase their reproductive success.

Most of our understanding of the adaptive value of monogamy is derived from studies of long-lived species with perennial pair bonds and of species in which pairs commonly reunite year after year. Coulson (1966, 1972) showed that female Black-legged Kittiwakes (*Rissa tridactyla*), that retained the same mate, initiated breeding earlier and produced larger clutches than non-retaining females. Increased individual breeding success has been related to pair bonding and mate fidelity in other species as well, including Red-billed Gulls (*Larus novaehollandiae*; Mills 1973), Northern Gannets (*Sula bassana*; Nelson 1966), Parastic Jaegers (*Stercorarius parasiticus*; Davis 1976), Manx Shearwaters (*Puffinus puffinus*; Brooke 1978), and Northern Fulmars (*Fulmarus glacialis*; Ollason and Dunnet 1978). Coulson (1966) argued that reproductive success reflected the compatibility of

---

[1] Present address: *Delta Waterfowl Research Station, RR 1, Portage la Prairie, Manitoba R1N 3A1, Canada.*

the individuals as cooperating parents which Erickson (1978) suggested might be established in part through mate choice. In birds that pair for life such as Zebra Finches (*Poephila guttata*), Lesser Snow Geese (*Anser c. caerulescens*), and Black-legged Kittiwakes, the original choice of a mate could have fitness consequences over several broods (Butterfield 1970; Cooke et al. 1982; Coulson 1966, 1972).

Pair-bonding is characteristic of mating systems in all tribes of Anatidae (Heinroth 1911; Delacour and Mayr 1945; Kear 1970). Early investigators studying captive or unmarked wild populations of ducks concluded that all species were monogamously pair bonded. However, recent studies have demonstrated that apparently monogamous mating systems are more complicated than they first appeared because they conceal a variety of secondary male reproductive tactics (McKinney et al. 1983; McKinney, Chap. 6). In the African Black Duck (*Anas sparsa*) paired males participate in extra-pair courtship leading to the formation of liaisons and, at times, subsequent mate changes (McKinney et al. 1978). In other ducks such as Mallards (*Anas platyrhynchos*), Pintails (*A. acuta*), and Green-winged Teal (*A. crecca*), paired males pursue other females and forcibly copulate with them (Smith 1968; Barrett 1973; Derrickson 1977; McKinney and Stolen 1982; McKinney et al. 1983). As an anti-cuckoldry tactic, males guard and actively defend their mates by attacking males attempting forced copulations (McKinney and Stolen 1982; McKinney et al. 1983). Females also make vigorous attempts to avoid forced copulations by hiding or escaping from their pursuers. During renesting periods, when forced copulations occur most frequently, paired males try to force copulations on their own mates (Burns et al. 1980; McKinney and Stolen 1982), which suggests that sperm competition may be important (Cheng et al. 1982, 1983). Thus, although monogamy is the primary mating system in Anatidae, the interests of individual males have resulted in a number of secondary reproductive tactics.

The process of pair-formation in wild waterfowl is poorly understood because few studies have used marked individuals. However, studies on individually marked captives indicate that in most waterfowl species, males perform courtship displays to females and females apparently choose among courting drakes (Weidmann 1956; Weidmann and Darley 1971). When the present study was initiated, little was known about mate preferences of Canvasbacks (*Aythya valisineria*). Therefore, the first objective was to examine the mate preferences of wild Canvasbacks and to determine whether female choice influences subsequent egg production. The hypothesis to be tested was that captive females placed with self-chosen mates would lay more eggs (and, presumably experience higher reproductive success) than females placed with randomly-assigned mates. The second objective was to describe the primary mating system of a flock of 150–250 captive Canvasbacks. The prediction here was that individuals in the flock would form monogamous pair bonds. This prediction was based on reviews of *Anas* breeding behavior (Weller 1964; McKinney 1975a) indicating that the basic mating system is seasonal monogamy. The third objective was to quantify the frequency of mate switching vs mate retention between years and to determine how this affects reproductive success of individuals. The prediction was that individuals retaining a given mate would lay more eggs than those that changed mates.

METHODS

The Canvasback breeding stock was reared from eggs collected in the field from 1973–1978 in southcentral Manitoba and Saskatchewan. The eggs were artificially incubated and hatched, and the birds were reared and maintained at the Delta Waterfowl Research Station (50°11′N, 98°19′W) using standard avicultural procedures (Ward and Batt 1973). All ducklings were web-clipped to identify siblings. Adult Canvasbacks were individually marked using numbered aluminum leg bands and numbered plastic nasal saddles (Bartonek and Dane 1964). Continuous records of age, health, and genealogy were maintained for all individuals.

The Canvasbacks were housed indoors over each winter (from early November until mid-April) and were isolated from other species of waterfowl. Adults (2 to 8 years old) over-wintered in 40–50 bird groups that were balanced for sexes and age classes. In mid-April the ducks were hand-carried from the wintering house and placed in an outdoor pen consisting of a concrete tank (30.5 × 12 × 0.4 m deep) covered by welded wire netting supported by a 1.8-m high steel frame. The tank had two equal compartments, each with a deep central trough. The compartments were filled with 30–40 cm of water. The large pen was subdivided into rows of 10–15 visually isolated cubicles and a central open area measuring 4 × 20 m.

Breeding cubicles measured 2.4 × 1.8 × 0.8 m and contained a metal feed pan, a concrete block protruding above the water surface for a loafing area, and a box filled with dried bulrush (*Scirpus* spp.) for nesting material. A standard commercial diet (Ward and Batt 1973) was provided ad libitum. The nest boxes (40.5 × 40.5 × 70 cm deep) were attached midway along the sides of the cubicles, 2 cm above the water surface. Ramps made of hardware cloth provided access to the nest boxes. Individual nest boxes were randomly placed in the open area of the pen where additional feeding and loafing ramps were also provided.

Observations were made of the wintering flock each of five springs (March–April 1976–1980) to identify the pairing status of the individual ducks. For the purpose of this paper, I define male-female social relationships in terms of pair-bonds, which I define as "mutual social attachments between mating partners that are strengthened by courtship and copulation." After four weeks of daily observations all individuals were classified in one of three categories according to their pairing status: (1) strong pairs (see criteria below), (2) actively displaying but not clearly formed pairs (males frequently performing displays during bouts of flock courtship and females Inciting (Hochbaum 1944) toward courting drakes), and (3) unpaired birds (males and females displaying very little). The criteria used to identify strong pairs were: (1) members of the dyad usually in close proximity (within 1–2 m), (2) the drake defending a particular hen's feeding territory around the communal food bowls, (3) a female consistently Inciting beside one particular male in a courting party, coupled with "chasing" other males away, and (4) attempted mounting and/or successful copulation of the pair, followed by simultaneous bathing.

Background on behavior in the wintering flock.—Repeated bouts of courtship by several males over at least several (3–5) days were required for pair formation. When captive Canvasbacks began courting in February and March, there was a period in which courtship consisted mainly of mutual exchanges of Neck-stretch displays among sexually active birds. Drakes also gave Head-throw and Kinked-neck displays (Hochbaum 1944). The frequency of the displays increased as the season progressed. By late March some females attracted several courting drakes simultaneously and "courting parties" of 3–8 drakes formed. Receptive females performed Inciting and usually approached one of the drakes in the courting party. Typically, these females gave Inciting beside several drakes for the first few days, thus increasing the frequencies of the drakes' approach behavior until one male eventually was able to maintain a position at the female's side. Strong pairs eventually developed from such associations. In most cases, the captive Canvasbacks formed monogamous pair bonds similar to those of wild birds (Hochbaum 1944; Weller 1965).

The importance of free mate choice.—I designed an experiment to test the importance of free mate choice on behavior and egg-laying. The controls consisted of one group of 19 strong pairs which were chosen randomly from 29 pairs in the main flock. These pairs were placed in randomly selected breeding cubicles. The remaining 10 pairs were separated from their mates and the females were assigned males that had been strongly paired to other females (Forced Pairs—Type I). That is, established strong pairs were broken up and their members assigned to other previously strong-pair individuals. A third group (Forced Pairs—Type II) consisted of 12 females that had been actively courted, but were not clearly paired. Each was placed in breeding cubicles with a mate chosen randomly from sexually active males that had not courted that particular female. All pairs were visually but not acoustically isolated from each other. All ducks used in this experiment were 2, 3, or 4 years old and only one female had previous breeding experience.

One week after pairs were placed in breeding cubicles, all were removed and placed in the open pen for two days. This was necessary because of the aggressive behavior of some females (Forced Pairs—Type I) toward the drakes placed with them. Observations of pairing behavior were made during these two days. Three of the 10 pairs that had initially been forced pairs (Type I) were reunited with their original mates. All other experimental dyads (the remaining 7 sets of Type I forced pairs and 12 pairs of Type II) were returned to their original breeding cubicles for the duration of the experiment. Nesting and egg-laying behavior were monitored for all pairs at 08:00, 11:30 and 14:00 h daily from mid-April through late June.

Results and Discussion

Importance of free mate choice.—The behavior of the Canvasbacks placed in breeding cubicles could be categorized clearly as one of three types—those in which (1) both male and female partners continued their courtship directed at each other; (2) females aggressively pecked and chased their male pen partners; and (3) males and females coexisted but did not court each other. Eighteen of the 19 control pairs continued their courtship activity. Generally these pairs adjusted rapidly to the cubicles. Most of these birds synchronized their activity patterns, fed, and gave courtship vocalizations within a few hours after being placed in the compartments. In sharp contrast, nine of 10 strongly paired females that were placed with males other than those they had chosen (Forced Pair—Type I) became

TABLE 1

NUMBERS OF EGG-LAYING FEMALES AND OF AGGRESSIVE FEMALES, AND DRAKE MORTALITY FOR THREE PAIR TYPES

Pair type	Description	Date	Number of aggressive females[1]	Drake mortality[2]	Number of egg-laying females[3]
Strong Pair (Controls)	Birds from strong, self-formed pairs	April 13–21	1/19	0/19	0/19
		April 23–June 30	1/19	0/19	17/19
Forced Pair (I)	Birds from strong self-formed pairs that had their mates exchanged	April 13–21	9/10	0/10	0/10
		April 23–June 30	6/7	5/7	0/7
Forced Pair (II)	Birds that courted actively but did not clearly pair. Courted but unpaired females were placed with randomly assigned drakes.	April 13–21	2/12	0/12	0/12
		April 23–June 30	2/12	0/12	0/12
Strong Pair	Birds from former Forced Pairs (I) (April 13–21) that were reunited with their self-chosen mates on April 23.	April 23–June 30	0/3	0/3	2/3

[1] $\chi^2 = 24.00$, $P < 0.01$ (d.f. = 2).
[2] $\chi^2 = 27.81$, $P < 0.01$ (d.f. = 3).
[3] $\chi^2 = 31.70$, $P < 0.01$ (d.f. = 3).

extremely aggressive toward their new companions ($\chi^2 = 24$, $P < 0.01$, d.f. = 2; Table 1). By the second day, these females actively chased and pecked the males whenever those males were in the water. Although males are larger and heavier than females, they did not retaliate. Under these conditions, five of seven drakes died from five to 45 days after introduction in the cubicles although no males in other experimental treatments died ($\chi^2 = 27.81$, $P < 0.01$, (d.f. = 3; Table 1). Among Forced Pairs—Type II, 10 of 12 dyads coexisted peacefully with each other but did not breed. These ducks did not exchange courtship displays, nor were females aggressive toward the drakes. Males and females did not synchronize their maintenance activities, as did strong pairs in the control group.

A significantly greater number of control females of strong pairs laid eggs than females of either forced pairs (Types I and II) ($\chi^2 = 31.7$, $P < 0.01$, d.f. = 3; Table 1). Seventeen of 19 females of strong self-formed pairs laid eggs while none of the females in the forced pairs laid. The average reproductive output was 0 eggs per forced pair vs 18.8 eggs per strong pair (Mann-Whitney U-test, $P < 0.001$). This high reproductive output of strong pairs continued for the remaining three years of the study, resulting in a total of 2600 eggs being produced (an average of 17.0 eggs per pair, 153 pairs).

Female Canvasbacks laid eggs only after they had formed a strong pair-bond and were isolated in breeding cubicles with that same drake. Because pair-bonding precedes reproduction in almost all ducks (Heinroth 1911; Kear 1970), and because stable preferences for individual mates are normal, the disruption of pair-bonds might be expected to inhibit egg laying. However, even courted but unpaired females did not form pair-bonds with drakes assigned to them without a normal social courtship and female-choice phase; such females laid no eggs in the following two and one-half months.

These results demonstrate that female Canvasbacks are highly discriminating and tenacious in their mate preferences. Female choice has been reported in a wide variety of animals (see reviews by Halliday 1978, 1983) including lek forming Sage Grouse (*Centrocercus urophasianus*; Wiley 1973), Ruff (*Philomachus pugnax*; Hogan-Warburg 1966; Rhijn 1973; Shepard 1975), Red-winged Blackbirds (*Agelaius phoeniceus*; Lenington 1980), a few ungulates (Uganda Kob (*Adenota kob*), Buechner and Schloeth 1965; Leuthold 1966), elephant seals (*Mirounga angustirostris*; Cox and LeBoeuf 1977), amphibians (see review by Arak 1983), fish (Semler 1971), and numerous insects (Borgia 1979; Thornhill 1980b; Parker 1979). However, none of these previous studies demonstrated that females may be tenacious in their mate preference once the initial choice was made. This may be due to the polygynous mating systems of the species involved and/or the low parental investment of the males.

In any species, the "quality" of the mate is presumably a critical determinant of reproductive success and the genetic constitution of the offspring (Halliday 1978, 1983; Partridge 1980). Therefore it is not surprising to find that animals do not pair indiscriminately. In many species, this selectivity depends upon refined perceptual discriminations and selective arousal of the animals involved (Burley 1981b). In this study, courted but unpaired females did not pair or lay eggs with randomly assigned mates, even when placed together for two and one-half months. Thus, female Canvasbacks seem to be more discriminating in their mate preferences than wild Mallards (*Anas platyrhynchos*) in captivity (pers. obs.) or domestic species such as Rock Doves (*Columba livia*; Klint and Enquist 1981), in which females will form pairs and lay fertile eggs when placed with randomly assigned mates. When given a free choice of mates, female ducks appear to regulate the pairing process (Johnsgard 1960; Weller 1965). They show interest or lack of interest in a drake by leading, following, attacking, or fleeing.

There is no obvious reason why mate choice is crucially important to female Canvasbacks but at least two possible explanations should be considered. First, it may be important for wild females in helping them secure feeding areas, defended by reliable mates, where they are free from harassment during the prelaying and laying periods (Anderson, Chap. 5). If males vary in their ability to provide this service, females might choose males through some criteria that accurately indicate the individual male's potential for defense. In other species of waterfowl, such as Gadwalls (*Anas strepera*; Dwyer 1974), Shovelers (*A. clypeata*; Afton 1979), and Bewick's Swans (*Cygnus columbianus bewickii*; Scott 1980), defense by the paired drake seems to provide the seclusion necessary for increased feeding by the female. Second, males may vary in their ability to locate feeding areas. Canvasback females rely heavily on sago pondweed tubers, snails, midge fly larvae, and other invertebrate foods during the prelaying and laying periods (Serie, pers. comm.). Because prairie potholes vary dramatically in their plant communities and invertebrate populations both temporally and spatially (Stoudt 1965, 1971), these preferred food supplies may be limited in their availability on certain potholes during the breeding season. Male Canvasbacks may differ in their ability to locate particularly rich potholes and females could gain by choosing "better" males.

Female aggressive behavior.—The dramatic aggressive behavior of females toward males other than their mates during the prelaying season appears to depend on their prior formation of a strong pair-bond: unpaired (but courted) females did not behave aggressively toward male penmates. Hochbaum (1944) noted that female Canvasback behavior in the wild changed temporally and that females became more "pugnacious" or "aggressive" toward courting drakes other than the preferred mate as the season progressed.

The literature on female aggressive behavior in birds is sparse (Gowaty 1981b), with most published studies concerning only intersexual or social dominance. Smith (1980) reported that female birds dominate their mates during the breeding season in 37 monogamous species from 18 families. For females, the early breeding season is a critical period in their annual cycle. At this time the female parental investment greatly exceeds that of males because they must bear the energetic costs of producing eggs, which are often quite large (Perrins 1970; Trivers 1972; Emlen and Oring 1977). Smith (1980) reasoned that if dominance permitted a female to obtain more food during this critical period, it could give her a clear advantage over more subordinate females. Indirectly, female aggressive behavior may also confer a selective advantage to their mates. A male that "allows" his mate to feed more during the prelaying period may father a larger and/or better-nourished clutch (Smith 1980) and hence gain by deferring his own feeding when food is limited.

I suggest that one important function of aggressive behavior in female Canvasbacks during the prelaying period is that of repelling courting drakes other than their mate. If female mate choice is adaptive, females should avoid forced copulations, which effectively cancels their choosiness (e.g., Gowaty, Chap. 2). Forced copulations have been observed in 39 species of Anatidae (McKinney et al. 1983). In wild Canvasbacks, attempts of forced copulations are rare, but "monogamous" male Canvasbacks may form extra liaisons after their mates begin to incubate (Anderson, Chap. 5). The aggressive behavior of paired females in the captive flock effectively kept other drakes away from the pair. These pairs were less likely to be disturbed by unpaired courting drakes during their mounting and/or copulation attempts. The aggressive behavior of females in the wild may also serve this function.

Individual recognition.—Immediately upon their release from the breeding cubicles on 21–23 April, the Canvasbacks joined the free-swimming flock in the open area of the breeding pen. The birds re-associated into their original pairing status groups (strongly paired, actively courted but not clearly paired, and unpaired). Fifty-eight birds that had been strongly paired (i.e., Strong Pairs and Forced Pairs—Type I) re-associated and formed 29 pairs. Another 24 birds (Forced Pairs—Type II) participated in courtship bouts but were not clearly paired. The remaining 20 birds in the flock were unpaired. Some members of the initial strong pairs re-associated with their original mates in less than 15 minutes and after two days, all 29 strong pairs had reformed. (See Appendix I for probability calculations.)

In this study, the mates chosen by the females reflected their own preferences in the sense that females did not choose mates assigned to them at random from the first two years that they were in captivity (1975–1976). In this experiment, Canvasbacks re-associated with their original mates after a separation of two days.

TABLE 2

AGE-RELATED PAIR FORMATION FROM 1976–1980[1]

Year	Number of pairs with mates of the same age	Number of pairs with mates of a different age	χ^2
1976	25	17	9.88*
1977	30	8	62.59**
1978	55	7	182.78***
1979	26	14	55.11**
1980	28	12	85.98***

[1] All Canvasbacks that chose mates in 1976 are recorded; the first year pair formation data were collected. After 1976, only Canvasbacks that did not retain their previous mates and chose new mates are recorded.
* $P < 0.05$. ** $P < 0.01$. *** $P < 0.001$.

Although this separation was brief, members of other pairs have re-associated after longer separation periods. I have seen members of three different pairs re-associate with their mates from the previous year after being physically separated in different groups over the winter (Nov. 10, 1977–March 15, 1978; Bluhm 1981). Members of one of these pairs, visually isolated but in auditory contact, copulated within 30 min of being placed together in the same flock in mid-March. Another captive pair has remained intact for seven years. Thus, captive Canvasbacks are capable of individual recognition and may show long-lasting mate preferences.

Age-related pair formation.—During the 1976–1980 period the occurrence of age-dependent pairing and mate retention between years was recorded. The Canvasbacks showed a significant tendency to choose mates of a similar age each year (Table 2).

Other studies have shown that birds use multiple criteria such as plumage color and pattern, courtship frequency, age, and reproductive experience in assessing the quality of a mate (Burley 1981b). Although I did not specifically test which criteria Canvasbacks were using to assess mate quality, the fact that they formed pairs according to age classes is provocative. Although other possibilities exist, one hypothesis is that birds of the same age may imprint sexually on each other while still ducklings or they may simply retain the ability to distinguish individuals they had been reared with from others in the mixed-age adult flock. Age-related pair formation has been documented in other avian species such as the Prairie Warbler (*Dendroica discolor*; Nolan 1978), Snow Geese (Cooke et al. 1976), and Canada Geese (*Branta canadensis maxima*; J. A. Cooper, pers. comm.).

In a subsample of 25 pairs of Canvasbacks of known genealogy, it was found that none of the birds had formed pairs with their siblings. This may be explained by the chronology of rearing procedures used at the Delta Station where newly hatched Canvasbacks were placed with their siblings only for the first 2–3 weeks. Afterwards, ducklings from a few (2–3) individual clutches were merged. At approximately two months of age, all ducklings were placed together and kept separate from adult Canvasbacks until the following May. Avoidance of sibling mating may have occurred as a result of early filial imprinting (when the ducks were kept in family groups) with sexual imprinting occurring later. The details of imprinting in Canvasbacks are not understood sufficiently to evaluate this possibility, but results reported here are consistent with Bateson's (1980) suggestion that imprinting enables an animal to learn the characteristics of its close kin so

TABLE 3

PAIR FORMATION 1977–1980	
Total number of pairs	167
Monogamous pairs	138 (82.6%)
Polygynous matings	29 (17.4%)
Polyandrous matings	0 (0.0%)

MALE PAIR FORMATION 1977–1980

Year	Total number of pairs	% of males	Monogamous pairs	% pairs	Polygynous matings	% matings
1977	43	(51%)	37	(86.0%)	6	(14.0%)
1978	40	(59%)	38	(95.0%)	2	(5.0%)
1979	47	(54%)	34	(72.3%)	13	(27.7%)
1980	37	(50%)	29	(78.4%)	8	(21.6%)

FEMALE PAIR INFORMATION 1977–1980

Year	Total number of pairs	% of females	Monogamous pairs	% pairs	Females mated with polygynous males	% of matings
1977	42	(62%)	37	(88.0%)	5	(11.9%)
1978	47	(70%)	40	(85.1%)	7	(14.9%)
1979	62	(73%)	34	(54.8%)	28	(45.2%)
1980	40	(66%)	24	(60.0%)	16	(40.0%)

that later it can choose a mate with slightly different phenotype (but not too different) from its parents and siblings. Such behavior may be the result of putative evolutionary pressures to avoid inbreeding on the one hand, and excessive outbreeding on the other (Mather 1943; Bateson 1978, 1980, 1982; Maynard Smith 1978; Shields 1982). This assumes a correlation between relatedness and similarity of external appearance. Recently, Bateson et al. (1980) have provided good evidence to support this assumption in Bewick's Swans.

Mating patterns.—Pair formation was observed in the free-swimming flock from March–April for four consecutive years. The majority of Canvasbacks were monogamous (Table 3). Fifty-three percent of the total number of Canvasbacks formed pairs with the majority (82.6%) being monogamous and the rest (17.4%) being polygynous.

The total number of males mating with females ranged from 50–59% of all the males in the flock and the majority of these were monogamous. Some of the males mated with more than one female. Generally, polygynous males mated with two females; however, one male paired with five females. This particular male's mating pattern accounted for the increase in polygyny during 1979. Although he copulated with all five females, he directed most of his courtship activity toward only two of them. Males involved in polygynous matings changed from year to year.

Because of the occurrence of polygyny, a higher proportion of females (62–73%) formed pair bonds than males, despite an equal sex-ratio in the flock. Most females mated monogamously: no females were polyandrous.

Only one forced copulation was observed in the flock during the four years of observations. The female in this situation was soliciting a copulation from her mate, but he lost his footing and fell off her back while attempting to mount. Immediately, an unpaired drake mounted the female and successfully copulated

TABLE 4

INCIDENCE OF CANVASBACK FEMALES THAT MATED WITH THE SAME OR DIFFERENT DRAKES

Years	Same mate		Different mate	
	(#)	(%)	(#)	(%)
1976–1977	4	16	21	84
1977–1978	10	18	45	82
1978–1979	29	49	30	51
1979–1980	19	34	36	66

with her (as indicated by her post-copulatory bathing movements). The female then rejoined her mate, as if unaware (or indifferent to the fact) that he had not copulated with her.

In recent years the focus on avian sociosexual interactions has shifted from the relationship itself to its fitness payoffs to each participant. Because male ducks apparently contribute very little to parental care (Ashcroft 1976; Sayler and Afton 1981), they are expected to seek additional mating opportunities. Recent work on dabbling ducks has demonstrated the existence of a variety of secondary male reproductive strategies (McKinney et al. 1983; McKinney, Chap. 6). These results show the main secondary tactic used by some of the captive males is polygyny. Presumably captivity provides "optimal" conditions for the economical defense of many females, and males are able to pair simultaneously with more than one female. In the wild, most males form a single bond early in the season. Later, a small number of males form liaisons with a second female when their original mate is incubating (Anderson, Chap. 5). Because forced copulations were rare in the captive Canvasbacks (as in the wild), such does not appear to be an important male reproductive tactic in this species.

Mate retention.—Mate retention between breeding seasons varied from 16–49% over the four years (Table 4). Mate retention did not correlate with previous breeding season success; in other words, birds that laid eggs did not re-pair preferentially with the same mate the following year. The majority (74%) of the pair bonds formed in the flock lasted only one breeding season. Fourteen percent of the flock retained the same mate for two years, 7% for three years, and 11% for three years or longer (Table 5). One pair remained together for seven years. This female laid fertile eggs every year and this pair was one of the highest egg-producers in the flock.

TABLE 5

DURATION OF MATE RETENTION

Years	Number of pairs
1	149 (74.0%)
2	29 (14.0%)
3	14 (7.0%)
4	5 (2.0%)
5	2 (1.0%)
6	1 (0.5%)
7	1 (0.5%)

Theoretically, both members of wild pairs might increase their reproductive success as a result of being strongly paired. However, because Canvasbacks are migratory, and males leave the females during incubation, it is unlikely that members of a pair would encounter each other during fall migration, or on the wintering grounds and re-form their previous association. This appears to be confirmed by field data. Mate retention between years in wild Canvasbacks is 0% (Anderson, Chap. 5). Furthermore, only female Canvasbacks are highly philopatric; 76% of adult females returned to their natal areas contrasted with 9% of adult males (Anderson, Chap. 5). In general, migratory ducks do not appear to form pair-bonds with the same mate for more than one season. Instead, they form pair-bonds on the wintering grounds, an act which may increase nesting success by facilitating earlier nesting (Lack 1968). A paired female may benefit from a male's presence by gaining increased feeding efficiency and dominance status. Once they arrive on the northern breeding grounds, female Canvasbacks probably form new pair-bonds only if their original mate dies or vanishes. Anderson (Chap. 5) observed one female switching mates after her original mate had been severely injured during the prelaying period. The female and her injured mate remained together for a few days, but she eventually left him and re-paired. Thus, in the wild, Canvasbacks appear to be tenacious in their pairing, after the original choice between mates has occurred.

Conclusions

The free choice of a mate and the process of pair formation is an important determinant of an individual Canvasback's reproductive success. In this study, variation in reproductive success was a result of large differences in egg production. Long-term observational studies of individually marked wild birds are needed to assess how reproductive success varies with the timing of pair formation and pair-bond duration. Studies of mating systems of tropical and southern hemisphere waterfowl species, which are suspected to have long-term pair-bonds, may provide valuable information on this point. We also need to determine how each component of reproductive success, such as clutch size, fledgling survival, and offspring survival to breeding age, varies with the age and experience of individuals and the duration of their pair-bond. Last of all, studies are needed to determine the inherited basis of differences in reproductive success.

Acknowledgments

My gratitude is extended to D. Mock, P. A. Gowaty, K. Cheng, M. Moore, M. Anderson, L. Oring and an anonymous reviewer for their suggestions and comments on this manuscript. I also thank my advisor, R. E. Phillips, and F. McKinney and M. Weller for their suggestions on earlier drafts of this manuscript, when it was a chapter in my Ph.D. dissertation. I thank F. Martin for his advice on experimental designs and statistical analyses. P. Ward, B. D. J. Batt, and E. K. Ward of the Delta Waterfowl Research Station are thanked for their assistance throughout the duration of this study. Last of all, I thank G. Collins and K. Kostow for their expert technical assistance in gathering behavioral observations on the captive flock. This research was funded by the Delta Waterfowl Research Station, the Ducks Unlimited Foundation, and the Agricultural Experiment Station of the University of Minnesota.

APPENDIX I

The null hypothesis was that all combinations of associations of birds from the three pairing status groups were equally likely (i.e., a strongly paired male might associate with an unpaired female, etc.). The actual probability equation is:

$$P = \frac{S!I!}{N!} \qquad \text{(Conover 1971)}$$

where P = observed probability
S = number of birds (in terms of pairs) showing a mate preference
I = number of birds showing no clear mate preference divided by two (= pairs)
N = total number of pairs in all treatments

Thus, for the observed probability of re-associating in original pairing status groups:

$$P = \frac{29!22!}{51!} = 6.4 \times 10^{-15} \quad \text{or} \quad P < 0.001$$

The probability of ducks re-associating with their original mate is represented by:

$$P = \frac{O!}{N!}$$

where P = observed probability
O = number of birds (in terms of pairs) associating with their original mate, and
N = total number of pairs in all treatments

Thus, for the observed probability of re-associating with a former mate

$$P = \frac{22!}{51!} = 7.3 \times 10^{-46} \quad \text{or} \quad P < 0.001$$

CHAPTER 5

VARIATIONS ON MONOGAMY IN CANVASBACKS (*AYTHYA VALISINERIA*)

MICHAEL G. ANDERSON

Delta Waterfowl Research Station, RR 1, Portage la Prairie, Manitoba R1N 3A1, Canada and Bell Museum of Natural History, University of Minnesota Minneapolis, MN 55455

ABSTRACT.—Observations of individually-marked wild canvasbacks (*Aythya valisineria*) in the parklands of southwestern Manitoba indicate that seasonal monogamy is their basic mating system. However, males commonly engaged in extra-pair courtship and sometimes bred with a second female after leaving their primary mate. Forced copulations and polygyny were also recorded, but seem to be very rare. Present evidence suggests that extra-pair courtship in canvasbacks is part of a secondary male reproductive strategy of serial monogamy. Comparisons of canvasbacks with other ducks suggest that both ecological and phylogenetic factors may have affected the evolution of these mating patterns.

INTRODUCTION

Monogamy is by far the most common mating system in birds (Lack 1968; Oring 1982). Despite this, most research on mating systems has focused on supposedly "more interesting" non-monogamous species. Consequently our understanding of the ecological and phylogenetic forces that have shaped monogamous systems in birds is limited.

Early investigators who studied captive or unmarked populations of ducks believed them to be basically monogamous (e.g., Heinroth 1911; Hochbaum 1944; Delacour 1954–1964). However, the assumption that ducks are monogamous raises a paradox: because male ducks ostensibly contribute very little parental care, theory predicts that they should be seeking additional mating opportunities (Trivers 1972; Emlen and Oring 1977; Maynard Smith 1977). If they are not doing so, we must ask, "Why not?"

Recent work on dabbling ducks (*Anas*) has shown that overlying a basic monogamous system in many species is a secondary male strategy of forcibly copulating with other fertilizable females (Smith 1968; Barash 1977; Derrickson 1977; Bailey et al. 1978; McKinney et al. 1978, 1983; Wishart and Knapton 1978; Burns et al. 1980; Cheng et al. 1982; McKinney, Chap. 6). In some species, pair males also engage in courtship of females other than their mates (e.g., McKinney and Stolen 1982).

No detailed behavioral study of a marked wild population of any pochard (*Aythya*) was available as I began my study, but with the dabbling duck experience as a backdrop, it seemed likely to me that in this group too, mating patterns might be more complicated than they appeared on the surface. The pochards are a rather homogenous assemblage of 15 species of freshwater diving ducks thought to be closely allied to the dabbling ducks (Johnsgard 1968) and thus a logical group to extend comparisons of social behavior.

METHODS

This study of mating patterns was part of a broader 5-year study of Canvasback (*Aythya valisineria*) social behavior (Anderson 1985). Studies of breeding birds were conducted primarily on a 31 km^2 area centered 9.6 km southeast of Minnedosa, Manitoba, Canada (50°10′N, 99°47′W). The native aspen parkland community of the area has largely been replaced by intensive small-grain farming. Knob and kettle topography has produced a high density of small wetlands numbering 26/km^2 and averaging 0.45 ha in size (Stoudt 1982). The area has a 30-year history of waterfowl studies and long-term information is available on population and habitat trends for most species, including the Canvasback (Olson 1964; Kiel et al. 1972; Trauger and Stoudt 1978; Stoudt 1982).

Behavioral observations were made during approximately 2200 h (myself) plus 400 h (assistants) from late-April to late-June in 1975 through 1979. Nearly all observations were made on 373 individually marked birds that were captured by one of several methods. Live-female decoy trapping (Anderson et al. 1980) provided the primary technique for capturing mature breeding birds in spring (51 males and 50 females). Additional incubating females (7) were caught using a modified nest trap (Weller 1957). Flightless immature Canvasbacks (137 males and 123 females) were caught by drive-trapping (Cowan and Hatter 1952) or night-lighting (Cummings and Hewitt 1964) between mid-July and mid-August 1975, 1976, and 1978. Five birds resident on the study area had been previously marked by U.S. Fish & Wildlife Service personnel. A few additional birds were marked with experimental dyes early in the study.

Captured birds were banded with standard USFWS aluminum bands and fitted with plastic nasal saddles (modified from Sugden and Poston [1968] and Doty and Greenwood [1974]), which were individually distinguishable by color and alpha-numeric code. Adults were examined for signs of breeding status (down pulled, pubis spread, egg in oviduct). Adults more than one year old were distinguished from ducks in their first spring using feather pattern criteria (Serie et al. 1982). Immatures were sexed by cloacal examination (Hochbaum 1942).

Systematic observations consisted mainly of intensively following a few marked pairs concurrently. Additional opportunistic or deliberate observations were also made as needed, e.g., to find nests or to ascertain pairing statuses at crucial times. Data reported here on pair-bonds and social interactions were collected during randomized sampling of marked pairs for a broad spectrum of behaviors and these methods are detailed elsewhere (Anderson 1985).

Details of social organization emerge grudingly from field studies. In only 47 cases of 115 marked-bird breeding sequences (one or both mates marked) was it possible to learn pair-bond histories with certainty. The data reported here involve only those 47 well-known pairs.

For data analyses the breeding season was divided into the following intervals: "Post-arrival"—first day on the breeding grounds to the beginning of nest-site exploration; "Pre-laying"—beginning of nest-site exploration to the day before laying of the first egg; "Laying"—laying period for the first clutch; "Incubation"—day of laying of the last egg to hatch or destruction of the first nest; "Renest Interval"—loss of an initial clutch to the day before laying the first egg of the second clutch; "Second Laying"—laying period for the second clutch; "Second

Incubation"—day of laying of the last egg to hatching or destruction of the second nest.

For this analysis, courtship time included all sexual displays and brief (<15 sec) pauses between displays. Data on courtship time were arc-sine transformed from percentage data and analyzed by single-classification analysis of variance (Sokal and Rohlf 1981). Because repeat observations were made on marked birds, data for different breeding statuses were not strictly independent. I could not analyze for individual bird effects using a split-plot design because all birds did not go through all breeding stages in all years and so could not be assigned randomly to different classification groups. Thus I had to assume these data were independent and analyzed them using a simple ANOVA design. This is a reasonable assumption as tests run for trends in the courtship data did not reveal any interdependence of errors ($P > 0.25$).

Results

Of 115 observed consort relationships among marked birds, 114 were one-male/one-female pairs, clearly suggesting that the Canvasbacks at Minnedosa were monogamous.

Philopatry.—Pairs formed on spring migration (Weller 1965; Anderson, pers. obs.) and males followed their mates back to traditional breeding grounds. By the time birds arrived at Minnedosa, 97% (N = 110) of returning resident females were paired. Females were highly philopatric and on average 76% of marked adult females and 27% of immature females returned each season. Evidence suggests that nearly all living females homed (Anderson 1985).

In contrast, among males only 9% of adults (5/52) and 1.5% of immatures (3/206) returned. Of these 8, all but 1 (an immature male) were seen only the year immediately after marking. No migrating males ever returned to the study area more than once. No repeat pairings were recorded among marked birds, and all homing males, regardless of their previous status, returned as unpaired birds. Of the 5 adult males that returned, 2 had been unpaired during the previous year and 3 had been paired. There was no evidence of mate retention for more than a single breeding season.

One-male/two-female trios.—I discovered only 1 polygynous trio in which a male maintained a simultaneous relationship with 2 females and copulated with both. The trio emerged late in the season when both females were in renesting intervals and many males had already left the study area. One trio female lost her renest clutch during laying. I was unable to determine the fate of the second female. The trio male was an unusually dominant and aggressive male and divided his attention between the two females, although one was decidedly favored. He attended her more closely approximately 70% of the time (N = 33 h) and responded more consistently to her moves and displays. Aggression between the two females was evident whenever the trio was together.

I observed 3 other trios which, upon casual inspection, would probably have been mistaken for polygynous trios. These occurred late in the season when 3 older renesting pairs were joined in their daily travels by post-breeding yearling daughters of the pair female. In these situations pair males actively displayed to

TABLE 1

COURTSHIP BY PAIRED MALES OF FEMALES OTHER THAN THEIR MATES (N = 24 MALES SAMPLED FOR AT LEAST 3 CONSECUTIVE BREEDING STAGES).[1] SEE TEXT FOR EXPLANATION OF BREEDING STATUSES

Breeding status of male's mate	Hours of observation	% time (s.e.)
Post-arrival	(138)	2.6 (0.6)
Pre-laying	(137)	2.2 (0.4)
Laying	(113)	1.2 (0.3)
Incubation	(92)	6.1 (0.2)
Renest interval	(71)	0.8 (0.4)
Second laying	(67)	1.3 (0.1)
Second incubation	(54)	10.3 (0.4)

[1] Single-classification ANOVA, $F(6,664) = 6.58$, $P < 0.05$.

the "extra" females but were never observed copulating with them. None of these young females renested.

Forced copulations.—Only 4 forced copulation attempts (for criteria see McKinney et al. 1983) were recorded during this study (1 forced copulation for every 9 intensively-observed pairs or every 650 h of observation). This is in contrast to >100 normal pair copulations that I observed. Marked pairs normally copulated 2 or 3 times per day (Anderson 1985).

All 4 forced copulation attempts were made by paired males, 2 by males with incubating mates and 2 by males with post-breeding mates. No forced copulation attempts were recorded among unpaired males. Of the females with which males forcibly copulated, 1 was laying a first clutch, 1 was a post-breeding female, and 2 were of uncertain status (but either in renest intervals or second-laying periods). As in the 1 case of polygyny, 3 forced copulations occurred late in the breeding season. The 1 forced copulation recorded in mid-season was in a very dry year (1977) when no renesting occurred. The one polygynous male observed in this study also carried out 1 of these 4 forced copulations.

I recorded 2 forced pair copulation attempts (after McKinney et al. 1983). One immediately followed an attempted forced copulation by another male on the female. In neither case did the paired males pursue these attempted mountings of their resisting mates to completion.

Extra-pair courtship.—In contrast to polygyny and forced copulations, male courtship of females other than their mates was common. Virtually all males occasionally left their mate and displayed to other females. Males did this throughout the breeding season (Table 1), but especially when their own mates were incubating and seldom while their mates were either in a laying period or a renest interval.

Displaying males approached females of any status. However, laying females were approached significantly more often than expected and females in second-laying, incubation, and post-breeding condition were approached significantly less often than expected (Table 2). This was calculated by comparing the breeding statuses of females approached by displaying males vs the random probabilities of encountering females of that status (based on the proportion of observations of marked females of each status).

TABLE 2

APPROACHES FOR DISPLAY BY PAIRED MALES TO FEMALES OTHER THAN THEIR MATES. OBSERVED VALUES FOR EACH BREEDING STATUS ARE COMPARED WITH OVERALL PROBABILITIES OF ENCOUNTERING MARKED FEMALES OF EACH STATUS[1]

Breeding status of approached females	Observed	Expected	Contribution to χ^2
Post-arrival	114	94.6	1.8
Pre-laying	82	92.1	0.6
Laying	70	39.6	8.5
Incubation	12	10.5	0.1
Renest interval	54	55.9	0.1
Second laying	16	37.6	8.6
Second incubation	2	8.9	4.4
Post-breeding	2	12.8	7.8

[1] Overall $\chi^2 = 31.9$, 7 d.f., $P < 0.001$, N = 664 h.

However, because females in some breeding stages used ponds without conspecifics more frequently than at other times (Anderson 1985), it was necessary to examine the persistence of courtship directed at females of known reproductive status during those observations when females were on ponds with other Canvasbacks. This analysis shows that females in the presence of conspecifics were displayed to more extensively during renest intervals than during any other reproductive stage (Table 3). So, although displaying males approached females of all breeding stages, they were most persistent in displaying to renesting hens.

Mate switches.—Of the 47 marked-bird breeding sequences for which I knew the precise nature of individual mate relationships, 8 (17%) involved a mate switch sometime during the breeding season, and this pattern was roughly similar in each of the four years that I followed marked pairs (Table 4).

Of the 8 females that switched mates, 4 were in renesting intervals, 1 was in the pre-laying stage and 1 was of unknown status. The statuses of two others were not known precisely, but circumstantial evidence suggests that they too were in renest intervals. The single female that switched mates in pre-laying was originally paired with a male that injured a wing and subsequently was unable to fly between ponds.

TABLE 3

COURTSHIP OF MARKED FEMALES BY MALES OTHER THAN THEIR MATES (N = 14 FEMALES SAMPLED FOR AT LEAST 3 CONSECUTIVE BREEDING STAGES)[1]

Breeding status of female	Hours of observation	% time (s.e.)
Post-arrival	(85)	2.8 (0.4)
Pre-laying	(84)	3.1 (0.6)
Laying[2]	(109)	4.1 (1.1)
Incubation[2]	(44)	1.0 (0.1)
Renest interval	(90)	13.8 (1.2)

[1] Single-classification ANOVA, $F(4,407) = 5.86$, $P < 0.01$.

[2] First nests and renests were combined because of small samples of second-nesting observations. Females with first nests were courted slightly, but not significantly more than females with second nests.

TABLE 4

MATE SWITCHES AMONG MARKED CANVASBACK PAIRS

Year	Total pairs	Fates known	Switches (%)
1975	8	5	1 (20)
1976	22	12	3 (25)
1977	46	17	2 (12)
1978	39	13	2 (15)
Total	115	47	8 (17)

Of the 8 males that succeeded in acquiring mates during these switches, 3 were previously-paired males whose first mates were incubating initial clutches. Three other males were unmarked and their statuses unknown. However, based on their behavior and the statuses of most marked males on the study area at that time, it is likely that they too were mates of other incubating females. No switching males were ever observed subsequently attending their original mates. No switching males abandoned their first mates until those females were about 7 d into incubation. So, at least some females changed mates within a breeding season, and at least some paired males sequentially bred with more than one partner.

Two males that took over switching females were previously unpaired. Unpaired males were transient on the study area but spent much of their time moving about, displaying, approaching pairs, and apparently "testing" existing pair-bonds (Anderson, pers. obs.). It appears that occasionally this behavior leads to a breeding opportunity.

With so few well-documented mate changes it is difficult to suggest why certain pairs switched. In one instance, a renesting female switched after her first mate had abandoned her during incubation to establish a breeding relationship with an earlier renesting female. In one instance noted above, an original paired male was injured. In 2 of the remaining 3 cases involving marked original males, the females were observed actively rejecting (chasing, Inciting) their "old" mates during social courtship. In these 2 cases, the male and female had been farther apart during observations from arrival to egg-laying than were mates in a sample of pairs that did not later switch mates (Table 5). One such female was seen repeatedly without her mate prior to nesting, a pattern never observed in any other marked pair. Males of switching pairs initiated fewer mutual displays and showed less consistent activity coordination with their mates than did non-switching males (Table 5). The reverse was true for females (Table 5). The males in these 2 pairs displaced other males in agonistic encounters slightly less often than did males of non-switching pairs (53% vs 59%). However, the overall frequency of winning, losing, or drawing (defined by supplanting) during encounters with other Canvasbacks, prior to renesting, was independent of eventual mate switching ($G = 2.1$, 2 d.f., $P > 0.10$, $N = 463$ interactions). For similar aggressive encounters with other species, males of pairs that later switched mates won only 35% of the time compared with 70% for males of non-switching pairs ($G = 8.9$, 2 d.f., $P < 0.05$, $N = 344$).

DISCUSSION

Single-season monogamy was clearly the predominant mating system among Canvasbacks on the Minnedosa study area. Arguments concerning factors leading

TABLE 5

COMPARISONS OF PAIR BEHAVIOR BETWEEN 2 PAIRS THAT SWITCHED MATES AND 4 RANDOMLY-CHOSEN PAIRS THAT DID NOT. DATA FROM ARRIVAL TO EGG-LAYING[1]

	Switching pairs (s.d.)	Non-switching pairs (s.d.)
Mean distance between mates (m)	5.4 (4.1)	3.1 (1.2)**
Initiation of mutual displays[2]		
Males	0.9 (0.7)	1.2 (0.5)*
Females	1.5 (1.1)	1.3 (0.7)*
Activity coordination[3]		
Males	−0.5 (0.7)	+0.1 (0.2)*
Females	−1.2 (0.9)	−2.2 (0.9)*

[1] Means compared by t-tests, N = 177 h.
[2] Display bouts per hour.
[3] Net frequency of activity changes (+ = yes, − = no) by one bird within 15 sec of its partner changing activities. High values indicate relatively greater effort at pair coordination (see Anderson 1984).
* = $P < 0.05$, ** = $P < 0.01$.

to monogamy among Canvasbacks have been offered elsewhere (Anderson 1985), and center on at least three important factors: (1) indirect male parental investment, (2) female philopatry and the timing of pair formation, and (3) a combination of the timing of female breeding requirements and resource distribution patterns. These factors are briefly summarized below.

Males are providing several kinds of indirect parental investment, i.e., investment via their mates which ultimately benefit offspring. Males are vigilant in the company of their mates, chase encroaching conspecifics, give calls warning of approaching predators, and chase parasitic Redheads (*Aythya americana*; Anderson 1985). Still, this is little male investment compared with, for example, many passerines or geese. Furthermore, at least some of this indirect parental investment is potentially shareable (Wittenberger 1979) among offspring of more than a single consort. Thus, other factors must prevent males from monopolizing additional females.

Females are highly philopatric. This, together with the timing of pair formation, severely limits a male's options. By pairing early in spring, females seem to gain advantages of priority access to critical resources (e.g., food), improved feeding time efficiency, and reduced disturbance (Anderson, pers. obs.). Females may also profit from having additional time to assess potential mates (McKinney 1986), and from the opportunity to choose mates from greater concentrations of flocked migrating males. Because females are limiting for males (male-biased sex-ratios and higher female parental investment), males are probably selected to pair, within limits, whenever females are ready (Anderson 1985). Because males can only follow a single female back to her natal area, they have few options. They must either pair early and follow a single homing female or remain unpaired on migration and attempt to establish single or multiple bonds with resident breeding females, a pattern that rarely succeeded.

Breeding was spread over about two months at Minnedosa and on average, 58% of Canvasback nests were unsuccessful (Stoudt 1982). The accompanying asynchrony of female breeding schedules and thus resource requirements, coupled with the patchy spatial distribution of the Canvasbacks' key resources and large home ranges, apparently made it impossible for males to monopolize multiple mates simultaneously (Anderson 1985). So, although much of a male's indirect parental investment is potentially shareable by more than one female, breeding schedules almost never allow it. Conversely, highly asynchronous female schedules did create sequential breeding opportunities for males.

Despite these constraints, variations on monogamy were found. Males were opportunistic and sought additional matings and females appeared to be continually assessing their mates. Forced copulations were uncommon in Canvasbacks at Minnedosa and occurred late in the season. It is problematic whether or not these resulted in fertilizations. Polygyny was observed only once, also late in the season. However, it is interesting that in captive flocks where sexually-active males may be scarce, and all breeding requisites are available in a small area, simultaneous mulitple bonds may occur (Bluhm, Chap. 4; pers. obs.). So, at least a low incidence of polygyny may be possible in other natural environments. Circumstantial evidence suggests the possible occurrence of polygyny (or kinship trios?) in both Common Pochards (*Aythya ferina*) and Tufted Ducks (*A. fuligula*) (Bezzel 1964; Reichholf 1964).

On the other hand, extra-pair courtship by males was widespread at Minnedosa, and descriptions of behavior in unmarked Canvasbacks elsewhere (Erickson 1948) seem similar. Males frequently approached females that had recently arrived on the study area. If males normally develop individual relationships with several females whose home ranges overlap with theirs, strange females should attract a male's attention. The discovery that courting males also preferentially approached females in their first laying period (but not the second) was surprising. Because no females were recorded switching breeding partners at that stage and forced copulations were rare, the reasons for such preferences are unclear.

Females in renesting intervals were not approached more often than expected by chance, but they were courted far more intensively and persistently than other females. This was also the period during which females were most likely to change mates. With little or no possibility of kleptogamy, the costs to an individual female or male of abandoning a partner to start a new breeding attempt should be lowest immediately after loss of a first clutch. Therefore, it is not surprising that most mate switches occurred during renest intervals. Also, because operational sex-ratios (Emlen and Oring 1977) were typically more skewed at that time than earlier in the season, females were in a position to be "choosy." The ratio of unattached males to available females increased by 40% from early May to early June (1.07 vs 1.39) (Anderson, pers. obs.). Paired male aggression and male efforts in affiliative behavior were also correspondingly higher than during pre-laying (Anderson 1984, 1985).

Considering the observed pattern of extra-pair courtship in males, the statuses of courted females, and the evidence from mate switches, I suggest a working hypothesis that extra-pair courtship in Canvasbacks is an evolved strategy for the acquisition of second pair bonds by already-paired males. Similar arguments have been made for the function of extra-pair courtship in several species of birds (Mock 1979; McKinney and Stolen 1982).

Canvasbacks are the first wild anatid studied in detail in which extra-pair courtship seems to be such an important paired male activity. Extra-pair courtship has been observed in Mallards (*Anas platyrhynchos*) during the non-breeding period (Weidmann 1956; Lebret 1961; Raitasuo 1964; von de Wall 1965), in wild African Black Ducks (*A. sparsa*; McKinney et al. 1978), and in captive Green-winged Teal (*A. crecca carolinensis*; McKinney and Stolen 1982), but the goals of such behavior have remained unclear. This is also the only detailed study of a non-territorial duck in which males seldom attempt to force copulations on other fertilizable females. The question of what factor(s) may have affected the cost/benefit of these alternative male reproductive strategies in ducks remains open. I offer the following comparisons as a tentative step toward an answer and to point out areas and species where further work is needed.

An instructive comparison with Canvasbacks is the Lesser Scaup (*Aythya affinis*), its best-studied congener. Female Canvasbacks initiate breeding early and, except in dry years, commonly renest if an early nest is destroyed (Doty et al. 1984; Anderson 1985). This means that the cultivation of other liaisons by male Canvasbacks may pay off in a second breeding sequence provided that some mate switches occur between nesting attempts. Scaup initiate breeding later at the same latitude, seldom renest except in exceptional years (Rogers 1959, 1964; Afton 1984), and (perhaps consequently?) exhibit a high frequency of forced copulations (Hammell 1973; Afton 1983). Little clear-cut evidence is available for other *Aythya* species based on marked birds. Redheads seem to engage occasionally in both forced copulations and extra-pair courtship (Sayler, pers. comm.; pers. obs.) but, because they are semi-parasitic (Weller 1959), they are especially difficult to study. Ring-necked Ducks (*Aythya collaris*) have been recorded participating in both forced copulations and mate switching (Hohman, pers. comm.). Common Pochards and Tufted Ducks may provide a Palearctic parallel with Canvasback/Lesser Scaup reproductive strategies (Hohn 1943; Bezzel 1964, 1969; Cramp and Simmons 1977), but hard evidence of breeding patterns based on marked birds is lacking. Almost nothing is known of the breeding behavior of the four species of white-eye ducks in the wild, although Ferruginous Ducks (*A. nyroca*) might engage in extra-pair courtship (Dement'ev and Gladkov 1952:491).

Comparisons with the better-studied dabbling ducks (*Anas* spp.) remain highly speculative but suggest some fruitful avenues for further research. Several *Anas* species, like the Canvasback, may renest yet still combine pair-bonding with occasional forced copulations—Mallards, (McKinney et al. 1983 and references therein), Northern Pintails (*Anas acuta*; Smith 1968; Derrickson 1977), Blue-winged Teal (*A. discors*; Stewart and Titman 1980), and Green-winged Teal (McKinney and Stolen 1982). Why don't Canvasbacks attempt forced copulations more often?

Some fundamental differences between the two tribes may have predisposed Canvasbacks towards a different strategy. For example, specialization for diving has favored somewhat different body shapes between pochards and dabbling ducks. Consequently, pochards have somewhat heavier wing-loading (Raikow 1973; Humphrey and Livezey 1982), are somewhat less maneuverable, and perhaps aerial pursuits are more costly. At the same time, because of their specialized diving skills, female pochards may be relatively more difficult targets for forced copulations than dabbling ducks. In Northern Pintails (Derrickson 1977), and possibly other species (Titman and Seymour 1981; McKinney et al. 1983), females on land may have more difficulty escaping forced-copulation attempts than females that dive under water. However, at least male Lesser Scaup have overcome any such disadvantage.

Male Canvasbacks have significantly more time "free" from their mates than do male dabbling ducks (Anderson 1985). Even in late pre-laying and early laying periods, female Canvasbacks spend several hours each day working on their substantial over-water nests, which is much more than the brief time most dabbling duck hens seem to spend at dry-land nest scrapes. This female involvement leaves male Canvasbacks with more solitary time which might be used to cultivate liaisons with other females. Also, because Canvasbacks, like most pochards, are usually found in social groups throughout the breeding season (Mendall 1958; Bezzel 1969; Anderson 1985), they also have ready opportunities to interact repeatedly with several pairs and probably to develop individual relationships (including monitoring of breeding status) with many birds whose home ranges overlap theirs.

The patterns and probabilities of nesting mortality may also be a factor. From the limited data available, *Anas* hen mortality rates appear to be higher than for Canvasbacks (Sargeant 1972; Eberhardt and Sargeant 1977; Johnson and Sargeant 1977; Stoudt 1982). (Lesser Scaup hens are also upland nesters, and they too are frequently killed on nests [Afton 1983].) If this is generally true, then the relatively low chance of nesting mortality for Canvasback hens might favor cultivating liaisons with known individuals. Severe nesting hen mortality could make time-costly liaisons a less attractive option for males. Similarly, high rates of nest loss and high rates of renesting (both characteristic of many prairie-nesting *Anas* species) would lessen the probable payoff associated with each individual renesting liaison for many dabbling ducks compared with Canvasbacks.

Whatever the reason, it seems that for Canvasbacks the option of forcibly copulating with other females must produce a lower average fitness payoff than the option of establishing an exclusive second breeding relationship with a single female. However, the relative costs and benefits associated with extra-pair courtship and forced copulations have not been assessed for any single species, much less for broad inter-tribal comparisons. Because some ducks engage in both activities, e.g., Mallards (Humburg et al. 1978; Ohde et al. 1983; McKinney et al. 1983), further intraspecific comparisons of alternative male reproductive strategies seem especially promising for analyzing these trade-offs.

We certainly require studies of how secondary pair-bonds are cultivated and how females are targeted for forced-copulation attempts in several species. More long-term observational studies of wild birds are also badly needed to improve our ability to make detailed cross-species comparisons of variations on monogamy. We need studies concerning the subtle and indirect forms of male parental investment, its importance to females and subsequent offspring, and how male investment might limit male options outside the primary pair-bond. In Canvasbacks at least, female assessment of these male attributes might be a factor in mate switches. Lastly, we need much better information on the relationship of critical resources to spacing patterns and breeding synchrony in wild ducks, with the associated implications for understanding mating systems.

After 40 years of work on the mating systems of wild waterfowl, we are just beginning to understand some of the ecological forces and phylogenetic factors that have shaped those systems. The surprising complexity of waterfowl mating patterns, coupled with the diverse radiation of the family, offers an excellent system for general study of the evolution of subtly different mating options.

ACKNOWLEDGMENTS

Data presented here were collected during my tenure as a graduate student at the James Ford Bell Museum of Natural History, University of Minnesota, and at the Delta Waterfowl Research Station. Financial support was provided by the North American Wildlife Foundation through the Delta Waterfowl Research Station and by the Department of Ecology and Behavioral Biology, University of Minnesota. Special thanks are extended to Frank McKinney, Bruce Batt, and Peter Ward for their advice, support, and encouragement. Jerry Serie and the Northern Prairie Wildlife Research Center (USFWS) provided valuable assistance throughout the field work. My thanks go to the many field assistants who contributed to this aspect of the study, and especially Rod Sayler, Bill Hohman, and Dave White. I am indebted to Alan Afton, Bob Bailey, Rod Sayler, and many other Delta colleagues for countless arguments over pochard social systems and to Jerry Serie for freely sharing his knowledge of the Canvasback. Thanks also extend to Frank McKinney, Alan Afton, Diane Anderson, Patricia Gowaty, Doug Mock, Frank Rohwer, Jerry Serie, and Judy Rhymer for helpful comments on the manuscript.

CHAPTER 6

PRIMARY AND SECONDARY MALE REPRODUCTIVE STRATEGIES OF DABBLING DUCKS

Frank McKinney
Bell Museum of Natural History, University of Minnesota, Minneapolis, MN 55455

Abstract.—Monogamy is the primary mating system in dabbling ducks. Paired males contribute to their mate's breeding effort by (a) protecting the female from predators and disturbance by rival males while she feeds intensively during the period of egg production (probably all *Anas* species), (b) defending a territory within which the female feeds (several species), (c) helping to care for the ducklings (certain southern hemisphere species). Paired males also engage in forced extra-pair copulations (recorded in 21 of 37 *Anas* species to date) and in some species these may be important secondary reproductive strategies. Polygamy does not occur in holarctic species, probably because each male is unable to monopolize more than one female in synchronously breeding populations with male-biased adult sex-ratios, but bigamous behavior has been observed in captives of three southern hemisphere species (Cape Teal, *A. capensis*; Speckled Teal, *A. flavirostris*; White-cheeked Pintail, *A. bahamensis*). Opportunities for males to hold two mates simultaneously may occur in wild populations of such species because extended and/or irregular breeding seasons are likely to produce asynchrony in the breeding and molt schedules of individuals.

Introduction

In birds with monogamous mating systems, paired individuals are generally considered to have very limited possibilities for increasing their reproductive success by activities unrelated to their pair-bonds. Some instances of intraspecific nest parasitism may prove to benefit paired females by supplementing the eggs they incubate themselves, but this phenomenon has not yet been well documented (Yom-Tov 1980). Opportunistic extra-pair copulations and occasional polygyny have been observed in many species (Gladstone 1979; Oring 1982; Ford 1983; McKinney et al. 1984), but cost-benefit analyses in terms of male fitness have not been made. The prediction by Trivers (1972)—that males of monogamous species can be expected to try to inseminate additional females, without aiding or monopolizing them, and that "mixed male reproductive strategies" will result—seems logical but is still largely untested. There is circumstantial evidence for such mixed strategies in several avian species (Beecher and Beecher 1979; Mineau and Cooke 1979; Fujioka and Yamagishi 1981) and comparative studies of waterfowl (Anatidae) have suggested factors that appear to favor or prohibit their use (McKinney et al. 1983). My purpose here is to review evidence for secondary reproductive strategies in one group of primarily monogamous waterfowl and to suggest some promising lines for future research on such variations on the monogamy theme.

The dabbling ducks (tribe Anatini) comprise one of the most familiar and widely studied groups of birds. As surface-feeding ducks, they feed primarily by dabbling, head-dipping, and up-ending in shallow water. Their food consists of aquatic plants and invertebrates, strained from the water by lamellae on the sides of the bill. Fresh water is preferred, but some species use brackish lagoons and even sea shores. All spend time swimming but they walk well on land and come ashore to rest. Most nest-sites are on the ground in vegetation, but a few species use tree-holes. In all species incubation is by the female only. Usually ducklings are led, brooded, and protected by the female only, but the male helps in a few species.

The group has representatives on all continents except Antarctica. Most of the 37 living species (now usually placed in the single genus *Anas*) fall clearly into one of 7 sub-groups: mallards, wigeons, green-winged teals, pintails, and blue-winged ducks, found in both northern and southern hemispheres; and austral teals and silver teals occurring only in the southern hemisphere. Many geographic races have been distinguished, including 20 forms endemic to islands (Weller 1980). Most species (26) breed primarily in temperate, subarctic or alpine zones, and many of these make extensive seasonal migrations. Of the remaining 11 species breeding in subtropical and/or tropical regions, many are sedentary, but two are nomadic.

With such a worldwide distribution, the members of the genus *Anas* are well suited for comparative, correlational studies involving factors dependent on latitude, geography, climate, and ecology. In this review I stress the importance of factors likely to influence the temporal availability of females, because this is especially relevant to secondary male strategies.

Although much information is available on diverse aspects of the biology of many dabbling ducks (Sowls 1955; Johnsgard 1965, 1978; Frith 1967; Bellrose 1976; Palmer 1976; Cramp and Simmons 1977; Bookhout 1979; Todd 1979; Weller 1980) data on mating systems based on study of wild, marked birds is lacking for most species. Almost all species (and many races) have been kept and bred in zoos and private collections, but few studies on captives have focused on the variations in monogamy that concern us here. The traditional view has been that dabbling ducks are monogamous and little attention has been paid to extra-pair-bond activities of males. Extra-pair copulations have been observed in 21 *Anas* species, but only a few of these species have been studied thoroughly enough to permit speculation on the probable significance of such copulations in enhancing male fitness (reviewed in detail by McKinney et al. 1983). Bigamous behavior has been observed in only three species (all captives) and, because these findings are new, most attention is given to this topic here.

Seasonal Monogamy: The Basic *Anas* Mating System

In *Anas* species breeding in temperate and subarctic regions of the northern hemisphere, pair-bonds form on the wintering grounds, usually several months before breeding begins. Pairs migrate northward in spring, the female leading her mate back to the area where she bred in previous seasons. These pair-bonds break during the short breeding season (April–June), usually while the female is incubating. If early clutches are taken by predators, one or several replacement clutches may be laid. The same pair-bond may be retained for these renest attempts or mate-switches may occur. New pair-bonds form each winter and it is probably rare for individuals to breed together in successive seasons. Therefore, in these species the basic breeding system is seasonal monogamy and mate-choice has nothing to do with territory quality. (For reviews of *Anas* breeding behavior see Weller 1964; McKinney 1975a.)

The process of pair formation has not been studied in marked individuals of any wild *Anas* populations, but there are many clues to the basic mechanisms from studies of individually marked captives (Weidmann 1956; Weidmann and Darley 1971; McKinney 1975b; Cheng et al. 1978, 1979; Laurie-Ahlberg and McKinney 1979; Standen 1980; Williams 1983) and from field observations on unmarked birds (Lebret 1961; Wishart 1983). Males intent on pairing approach females and direct courtship displays toward them. Females respond to such males by threatening them, by showing acceptance of them (by performing inciting displays beside them), or by various kinds of behavior that encourage male courtship (e.g., nod-swimming). The adult sex ratio in wintering populations is usually skewed in favor of males (Bellrose et al. 1961), so that competition is intense and some males fail to secure a mate. When a number of males direct courtship toward the same female, "social courtship" results. This usually occurs on water. As well as directing displays at the female and maneuvering to attract and hold her attention, males threaten one another and perform aggressive displays. Chasing and fighting tend to be infrequent, but subordinate males perform appeasement displays (e.g., nod-swim) that seem to enable them to remain in the competition.

The female makes her initial choice from among the males that court her and indicates her preference by performing inciting displays beside the chosen male and threatening other males. When given free choice, females exhibit clear preferences (Cheng et al. 1978) apparently based on behavioral and morphological features (Bossema and Kruijt 1982; Kruijt et al. 1982). Early bonds seem unstable and are likely to be contested by rival males. Switches occur, apparently as a result of the chosen male being displaced by a dominant male which is then accepted by the female. Disputes between well-matched males are settled by fighting.

Pair-bonds are presumably reinforced by displays, by copulations, and by collaboration by mates in rejecting advances of rival males. Males perform characteristic displays directed at the female when mates come together (e.g., belly-preen, preen-dorsal, preen-behind-wing, bridling, and lateral dabbling) and in some species the female may respond with similar displays (e.g., preen-behind-wing). Pair copulations, preceded by mutual head-pumping, occur during the winter several months before egg-laying begins. At this time (e.g., September in Mallards, *Anas platyrhynchos*), gonads are regressed and spermatogenesis is not occurring (Höhn 1947; Johnson 1961, 1966); so apparently these copulations are part of the mate-testing and/or pair-bond-maintaining processes. Mates give combined display performances during encounters with other pairs or with courting males. In some species similar displays are given by both sexes (e.g., rapid chin-lifting in wigeons); more commonly, the female performs inciting while the male adopts a threat posture or swims away (followed by the inciting female) giving turn-back-of-head.

In summary, *Anas* pair-bonding involves expression of individual preferences by males (directing displays at certain females) and females (inciting beside one of the courting males). Acceptance of the male by the female must play an important role in preserving the bond but, when challenged, the male must be able to drive away all rivals, if necessary by fighting. Wittenberger (1979) has termed this type of mating system "female-defense monogamy," but the implication that males control the system by sequestering a female (Wittenberger and Tilson 1980) is misleading in neglecting the role of females in the process.

Pair-bonds of some tropical and southern hemisphere *Anas* species are suspected to be long-term rather than being formed anew each year (Weller 1968; Kear 1970; Siegfried 1974). Retention or reforming of old bonds may well be more feasible in sedentary populations than in migratory ones, and this tendency might also be favored by natural selection in species where mates cooperate in brood-care and/or where breeding seasons are irregular and of variable duration. Although there are a few records of prolonged bonds in Cape Teal (*A. capensis*; Siegfried et al. 1976) and African Black Duck (*A. sparsa*; McKinney et al. 1978) firm evidence on this point is scarce. I suspect that considerable variation exists in the length of pair-bonds in *Anas* populations with extended and/or irregular breeding seasons. Judging from the occurrence of courtship behavior before each breeding season and/or breeding attempt, bonds are frequently contested and mate-switches are probably common. This is quite different from the situation in swans, geese, and whistling ducks, where pair-bonds are characteristically long-term and often lifelong.

Review of Male Mate-Support Roles

Paired males appear to contribute to their mate's breeding effort in three ways: (a) by protecting the female directly (probably all *Anas* species), (b) by defending a territory within which the female feeds (several species), and (c) by helping to care for the ducklings (certain southern hemisphere species). Obviously these activities restrict males' abilities to exploit secondary mating options.

Protection of the mate.—Throughout the duration of the pair-bond the male associates closely with his mate most of the time (except when she is at the nest). From the male's point of view, close escorting of his mate is necessary if he is to minimize the chances of (a) losing his mate to another male or (b) having his mate inseminated by other males. The importance of the latter hazard is indicated by the especially close escorting and guarding of females by their mates during the period just before and during egg-laying. It is during this same period that females must increase their intake of invertebrate food for egg-production (Krapu 1979), which necessitates more time spent feeding (Dwyer 1975; Afton 1979; Stewart and Titman 1980; Titman 1981). Ashcroft (1976) has shown for Common Eiders (*Somateria mollissima*) that the vigilant behavior of the male allows the female to feed more efficiently, the male taking over the "lookout" role for both birds, and the same situation probably applies in dabbling ducks.

The role of male vigilance in protecting his mate from surprise attack by predators and from attempts by other males to force copulation is strongly indicated by observations on the behavior of pairs (Seymour and Titman 1978). Mates keep in close vocal contact with one another, giving very quiet calls as they move around together, loud contact calls when separated (decrescendo call, repeated calls). Females respond to alarm calls by their mates by stopping feeding and also becoming alert. Males also call to their mates while the latter are on their nests (Dwyer 1974). In some situations paired males crouch to avoid detection by flying males and move out of sight with the female.

Defense of a breeding territory.—Paired males of several *Anas* species defend stretches of water against intruding conspecifics for periods of about one week (Mallard; Dzubin 1969) up to several weeks (Northern Shoveler, *A. clypeata*; Poston 1969; Seymour 1974). Male Mallards hold territories at the beginning of the breeding season while their mates are in pre-laying and laying phases, but Shovelers continue to defend their territories throughout the incubation period. Provision of a secluded area, where females can feed efficiently while protected from interruption by their escorting mates, appears to be an important benefit of territoriality in the Northern Shoveler (McKinney 1973; Seymour 1974; Afton 1979) and probably in certain other territorial species (Gadwall, *A. strepera*; Dwyer 1974; Blue-winged Teal, *A. discors*; Stewart and Titman 1980).

African Black Duck pairs hold territories on rivers and in one study (Ball et al. 1978) territories were maintained throughout the year except for the period of wing-molt. In this case, the river territory could serve all needs of the pair (food for the pair and their ducklings, cover for nest-sites, and safe molting places). There was strong competition for mates and for territories, and only territory-owning pairs bred. Other species living on rivers (Salvadori's Duck, *A. waigiuensis*; Kear 1975) or shorelines (Crested Duck, *A. specularioides*; Weller 1972) hold similar all-purpose territories. Probably this system is favored in primarily insectivorous species by the need to control an exclusive area that will ensure an adequate food supply for the young in addition to other benefits.

White-cheeked Pintails (*A. bahamensis*) behave territorially, both in the wild and in flight pens, but male intolerance appears to be associated especially with defense of the mate against rival males and it is not clear if territories are important for female feeding seclusion (McKinney and Bruggers 1983). Other *Anas* species are known to be territorial (Brown Teal, *A. aucklandica chlorotis*; Hayes 1981; pers. obs.) but have not been studied intensively.

Some *Anas* species do not establish breeding territories. This is the case in the Northern Pintail (*A. acuta*), a species with large breeding pair ranges that may be economically undefendable (Derrickson 1978). Similarly, Green-winged Teal (*A. crecca*) and Red-billed Pintail (*A. erythrorhyncha*) are non-territorial (McKinney and Stolen 1982; pers. obs. on wild birds). Flight pen observations suggest that the same is true in Speckled Teal (*A. flavirostris*) and Brown Pintail (*A. georgica*).

Assistance in care of ducklings.—Males of all holarctic *Anas* species desert their mates before (or occasionally shortly after) the ducklings hatch and move to safe areas where they undergo the wing-molt. In the southern hemisphere, broods of many species are observed with two adults in attendance and in some cases males appear to assist their mates in protecting their offspring. However, for most southern species, documentation of male brood care in wild broods is poor because (a) the sex of the adults could not be determined (most of these species are monomorphic), (b) it was not known if the second adult was the female's mate, and/or (c) behavioral observations were not made. In some cases, records for broods raised in captivity confirm field reports of biparental attendance but descriptions of behavior observed are scarce.

Based on the fragmentary evidence available, brood care patterns in southern hemisphere species appear to be of three types: (a) male usually present and apparently contributes to protection of ducklings, (b) male usually not present, and (c) male sometimes present (Table 1).

TABLE 1

EVIDENCE ON PARENTAL ROLES IN BROOD-CARE IN SOUTHERN HEMISPHERE *Anas*

Brood-care categories	Number of wild broods seen with: 2 adults	1 adult	Category based on captives	Reference no.: wild	captives
A. Biparental care usual					
sibilatrix	4		A	1	2, 4
capensis	96	10	A	3, 4	4
specularioides	11		A	1	2
castanea	3		A	4	2
	>50%			5, 6	
aucklandica chlorotis	1		A	4	4, 10
B. Male's presence and/or care variable or uncertain					
flavirostris	3	2	B	1	4
gibberifrons gracilis		usual	A	5	2
a. aucklandica		?		7	
g. georgica	4	1		8, 15	
georgica spinicauda	1		B	1	4
b. bahamensis	5		B	9	4
bahamensis galapagensis		a few		10	
erythrorhyncha	24	55	B	3	4
specularis			A		11
versicolor	occurs		A	12	11, 15
waigiuensis	presumed		A	10	
C. Brood care by female only					
sparsa	4	33		3, 13	
s. superciliosa		usual		10, 4	
u. undulata	14	284		3	
hottentota	2	17	A	3	11
		usual		14	
smithii	33	225		3	
rhynchotis		usual		5	

Reference numbers: 1. Weller 1975a; 2. Kear 1970; 3. Siegfried 1974; 4. McKinney, unpublished; 5. Frith 1967; 6. F. I. Norman, pers. comm.; 7. Weller 1975b; 8. Weller 1975c; 9. McKinney and Bruggers 1983; 10. Johnsgard 1978; 11. M. Ounsted, pers. comm.; 12. Weller 1967; 13. McKinney et al. 1978; 14. Clark 1969; 15. Todd 1979.

I have observed wild broods of Cape Teal, Chestnut Teal (*A. castanea*), and Brown Teal and noted several kinds of male care. In all cases, males were extremely watchful and alert for predators, spending most of the time in head-erect postures while the ducklings fed. Male Chestnut Teal also gave warning calls, escorted lagging ducklings, and attacked hovering birds of prey. Male Crested Ducks, Bronze-winged Ducks (*A. specularis*), and Chiloe Wigeon (*A. sibilatrix*) are noted for their attentiveness to broods, and probably they provide similar kinds of protection. Most of these species have not been intensively studied in the field, and it is not known whether broods are especially vulnerable to predation. In the Cape Teal, this seems likely to be the case, because open habitats devoid of emergent cover are commonly used.

Two parents are not usually seen with broods of African Black Duck, African Yellowbill (*A. undulata*), Hottentot Teal (*A. hottentota*), Cape Shoveler (*A. smithii*), and Australasian Shoveler (*A. rhynchotis*). In two marked pairs of African Black Ducks, the female alone led their ducklings all day. At night, however, they joined the male at favored roosting places on their river territories, and it is possible that males help protect their broods from nocturnal predators.

In five southern species (Speckled Teal, Brown Pintail, White-cheeked Pintail,

Red-billed Pintail, and Grey Teal, *A. gibberifrons*) the relationship of males to their broods is poorly understood. Some broods are accompanied by two adults, others by just one. In some cases observers disagree on which is the typical pattern for the species. Active participation by males in care of ducklings has not been described. No intensive studies of wild broods have been made, but I have watched captives of the first four species and noted less interest by escorting males in the ducklings than was typical in Chestnut Teal and Cape Teal.

Three factors might contribute to the impression of "partial male attendance" in these species. First, the need for male participation in brood care could vary in different parts of the species' range (e.g., in relation to availability of escape cover or predator densities). Second, the accompanying male may be interested primarily in the female rather than the brood (e.g., maintaining his pair-bond for future breeding attempts, as suggested by Siegfried 1974). Third, males might be adopting a bet-hedging strategy, either escorting their broods (mate-holding and/or protecting ducklings) or deserting for an alternate mate when circumstances favor it.

Benefits of Monogamy to Males and Females

By pairing with a single female, a male can protect his genetic paternity and minimize the risk of cuckoldry. The importance of this factor in dabbling ducks is shown by (a) the increased vigilance of males while their females are fertile and (b) active defense of the female against males attempting to force copulation (p. 76). A male also stands to enhance his own reproductive success by providing vigilance, warning, and defense for his mate. Male vigilance during the fertile period is probably universal in *Anas,* but thereafter desertion of the female may occur at various stages (e.g., early, middle, or late incubation in holarctic species).

Females stand to benefit in various ways from having a male escort (protection from harassing males and predators), but the extent to which males contribute to their mate's breeding efforts is likely to change during the course of the breeding season. Anderson (1984a) has measured such changes in the Canvasback (*Aythya valisineria*), and there are probably many parallels with dabbling ducks. The interests of males and females are likely to coincide in promoting strong mate-defense by males during the pre-laying and laying phases. For males this is the crucial time for protection of paternal investment; for females it is a risky time when nest-sites are being selected, and when intensive feeding is needed to produce eggs. Later, during the incubation period, males are likely to have conflicts of interest. If the male remains in close contact with the incubating female, he will be in a good position to fertilize the eggs in her renest clutches, if these are required. On the other hand, the payoff may be greater if he deserts his mate and proceeds to a molting site in preparation for the wing-molt (e.g., if the earliest males to molt have advantages in successful migration, wintering or pairing). Alternatively, males may have opportunities to engage in secondary reproductive strategies. In regions where breeding is possible over an extended period, males may be better off staying with the same female throughout the brood-rearing period and keeping the same mate for subsequent breeding attempts. Here again, the interests of both partners (though not identical) may lead to persistence of the bond.

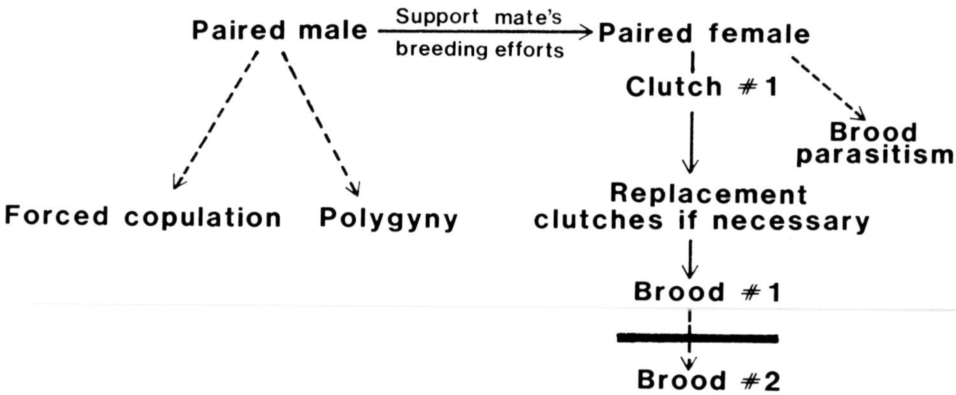

FIG. 1. Primary (solid lines) and secondary (dashed lines) breeding options theoretically available to *Anas* species. Two broods require a lengthy breeding season usually impossible in holarctic species.

Secondary Reproductive Strategies Combined with Monogamy

In addition to collaborating with their mating partners in jointly raising ducklings, both male and female dabbling ducks have other reproductive options (Fig. 1). For females, the major option would seem to be brood parasitism, laying in the nests of other birds. Although both inter- and intraspecific brood parasitism occur in a number of waterfowl (Weller 1959; Payne 1977; Yom-Tov 1980), this behavior appears to be rare in most species of dabbling ducks. Most *Anas* species place their nests on the ground, in vegetation, often far from water, and nests tend to be widely dispersed. Brood parasitism is common in waterfowl that nest over water (pochards, stiff-tails) or in tree-holes (perching ducks, whistling ducks), presumably because nests are more concentrated and easily located. Among dabbling ducks, it may be common only in a few species that frequently use tree-holes as nest-sites (e.g., Chestnut Teal; Norman 1982).

In most parts of the northern hemisphere, there is not sufficient time for female dabbling ducks to rear two broods each year and the annual cycle of these species is geared at most to producing several replacement clutches and rearing a single brood. In many tropical and southern hemisphere species, breeding can occur over a much longer period and double-broodedness seems to occur when conditions are favorable (Braithwaite, 1976b).

For males two possibilities exist for secondary reproductive strategies: extra-pair copulation and polygyny.

Forced copulation.—Dabbling ducks are well known for the occurrence of forced copulation (FC). Males vigorously chase females in aerial pursuits, females are overpowered and mounted (on land or water) and, especially when many males participate, the female's mate may be unable to prevent forced insemination. FC has been recorded in 21 *Anas* species, and probably it occurs in all but a few members of the group. (The only species in which it has been looked for and *not* found is the African Black Duck.) Evidence indicating that FC is a secondary reproductive strategy of paired males in certain *Anas* species has been reviewed in detail elsewhere (McKinney et al. 1983) and will be summarized only briefly here. The key points are:

1. Eggs can be fertilized by insemination during FC (Burns et al. 1980).
2. In all species studied intensively, FCs are performed primarily by paired males; unpaired males court females and try to obtain mates.
3. In the best studied holarctic species, FCs occur during the period when eggs are being fertilized, and in captive Mallards FC attempts were directed primarily at females in pre-laying and laying condition (Cheng et al. 1982).
4. Paired males usually defend their mates by attacking males that attempt FC.
5. Paired males may force copulation with their own mates. This often occurs shortly after their mates have been subjected to FC, suggesting an antidote insemination strategy.
6. Experiments using artificial insemination and genetic plumage markers in Mallards have shown that (a) the second of two competing inseminations 6 h apart overlays the former insemination and is 70% more potent, (2) there is an insemination "window" within 1 h of oviposition when the next egg in the clutch is fertilized (Cheng et al. 1983). The extent to which males time their copulations to take advantage of these temporal factors influencing sperm competition is not known.

The proportion of eggs in wild clutches that are fertilized by males other than the female's mate has not been determined for any *Anas* species. Also, judging from the frequency with which it is observed, FC is probably more important as an insemination technique in some species (Mallard, Northern Pintail) than in others (Northern Shoveler). In species such as the Northern Pintail, paired males undoubtedly spend time and energy seeking FC opportunities, and the concept of a "mixed male strategy" (Trivers 1972) is appropriate. In species such as the Northern Shoveler, where males defend a breeding territory and spend most of their time guarding it, FC is probably an option that males exploit only occasionally when the opportunity presents itself.

Polygyny.—There appears to be no firm evidence that polygyny occurs in wild populations of any holarctic *Anas* species, although successive monogamy can occur within one season (Humburg et al. 1978) and instances of bigamy have occurred when sex-ratios were experimentally altered in favor of females (Ohde et al. 1983). Typically these species have skewed sex-ratios, with a preponderance of males (Bellrose et al. 1961) and it is unlikely that a male could hold more than one mate at a time in unmanipulated populations. Recently, however, we have observed instances of bigamy in three southern hemisphere species under flight pen conditions. No such behavior occurred in three holarctic species (Mallard, Northern Shoveler, Green-winged Teal) breeding in the same pens in other years.

Bigamous relationships developed in 1981 and 1982 in groups of three pairs of Speckled Teal in a flight pen (27.5 m × 55.0 m × 3.6 m high). These birds were bred from stock obtained from aviculturists in the USA and they appeared to be typical representatives of the Chilean race (*Anas f. flavirostris*). The same male was involved in both years, but the females were different. The events were carefully documented in 1982 (Fig. 2).

During May and early June, male W was strongly paired to female W (as were the Y and R pairs). Pair W copulated regularly and male W also made many FC attempts on female Y, which was in wing-molt much of this time and especially vulnerable. On 19 June male W suddenly broke up the Y pair by dominating

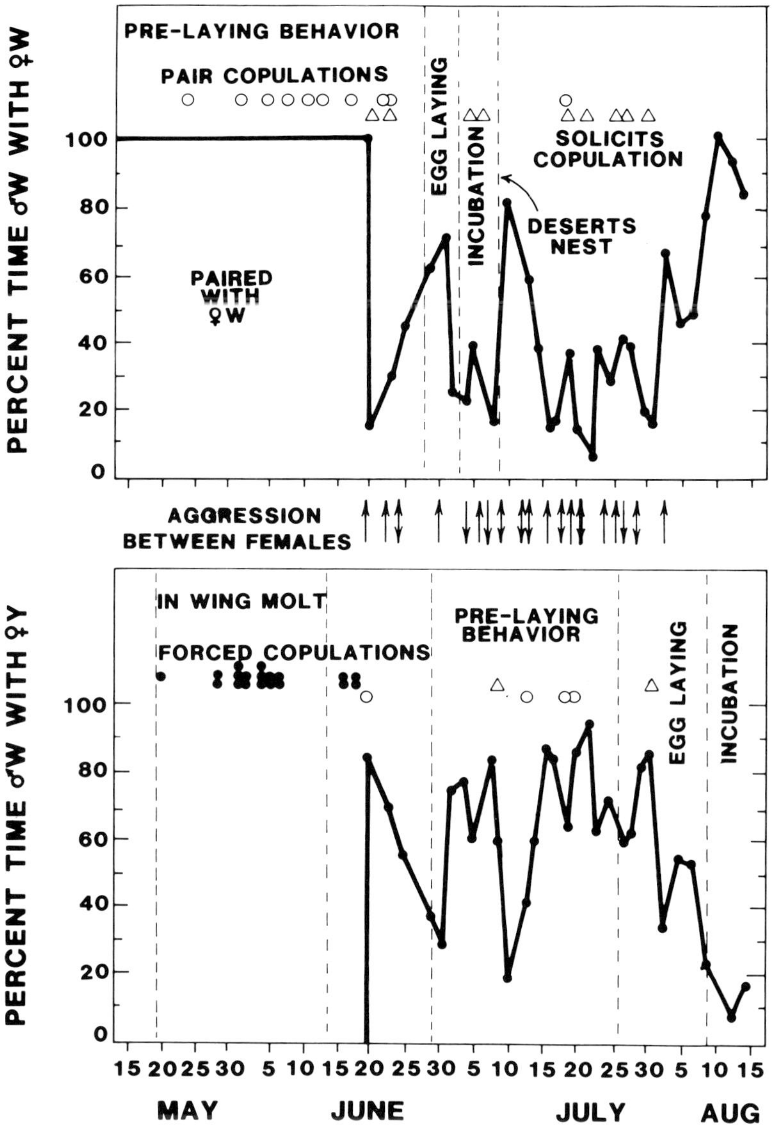

FIG. 2. Bigamous behavior of captive Speckled Teal male W with primary female W (above) and secondary female Y (below) in 1982. ● forced copulation attempts; △ pair copulations initiated by female; ○ pair copulations, initiator uncertain.

male Y and keeping him away from female Y. Female Y accepted male W as a mate and they were seen to perform a pair copulation. (Pair copulations are preceded by pre-copulatory displays and occur only when the female is willing.) On subsequent days, male W divided his time between the two females, repeatedly driving off male Y whenever he came close to female Y. Male W succeeded in maintaining this bigamous relationship for almost two months.

The proportion of time male W spent with each female changed at various times during the study period. He spent more time with female W while she was egg-laying, less time with her when she began to incubate, and again more time with her when she deserted her nest. Conversely, male W spent more time with female Y while she was in pre-laying and early laying phases, and less when she was incubating.

The two females showed frequent hostility toward one another, threatening, chasing, and at times fighting. Each female tried to monopolize male W, actively following him and inciting beside him. During periods when female W was being neglected, she initiated pre-copulatory behavior and actively solicited copulation from male W. Similar solicitation was also noted by female Y on two occasions; in both cases, this happened on days when male W was showing renewed interest in female W. We interpret this behavior as a mate-holding tactic. This is supported by the fact that female W gave such solicitations during her incubation period, a time when females are not expected to show any interest in copulation.

Very similar sequences of events occurred in cases of bigamy observed in Cape Teal (3 pairs of wild-caught birds in a flight pen measuring 27.5 m × 27.5 m × 3.6 m) (Stolen and McKinney 1983) and White-cheeked Pintail (3 pairs, probably the Caribbean race *A. b. bahamensis,* in the same pen as the Speckled Teal study) (McKinney and Bruggers 1983). In all cases, a paired male suddenly switched from making FC attempts on a second paired female to courting her and dominating her mate. Copulation solicitation and rivalry between females similar to that observed in Speckled Teal occurred in the Cape Teal.

These three species differ from holarctic dabbling ducks in having extended breeding seasons. Both in the wild and in captivity Cape Teal engage in year-round courtship (Johnsgard 1965; pers. obs.), and in many parts of the range breeding is irregular in response to unpredictable rains (Siegfried 1974; Winterbottom 1974; Dean 1978). Speckled Teal also have year-round courtship (von de Wall 1965; Standen 1976), and at least in parts of their range the breeding season is extended and possibly two broods are reared (Johnson 1965). In the southern Bahamas White-cheeked Pintails breed over a long period (McKinney and Bruggers 1983).

With extended and/or irregular breeding seasons, multiple broods are possible, and pronounced asynchrony in the reproductive condition of individuals within local populations is likely. A factor that could be very important here is the timing of the wing-molt. Flight feathers need to be replaced about once per year and waterfowl achieve this by a simultaneous molt after which flight is impossible for 3 or 4 weeks. In species with regular annual cycles the wing-molt occurs immediately after the breeding season (Bellrose 1976; Cramp and Simmons 1977) but with extended and/or irregular breeding seasons some individuals may be flightless while others are breeding (Braithwaite 1976b; Dean 1978). In some species, parents can become flightless while brood-rearing (e.g., Cape Teal; pers. obs.), but otherwise the wing-molt is temporally separated from breeding activities and flightless females offer no immediate reproductive possibilities for males.

No field studies have been made on Cape Teal, Speckled Teal, or White-cheeked Pintail with the objective of relating breeding seasons, gonad condition, and the timing of wing-molts. However, the studies by Frith (1959), Braithwaite and Frith (1969) and Braithwaite (1976a, b) provide such information for four Australasian *Anas* species. Chestnut Teal, Australasian Shoveler, and Australasian Black Duck

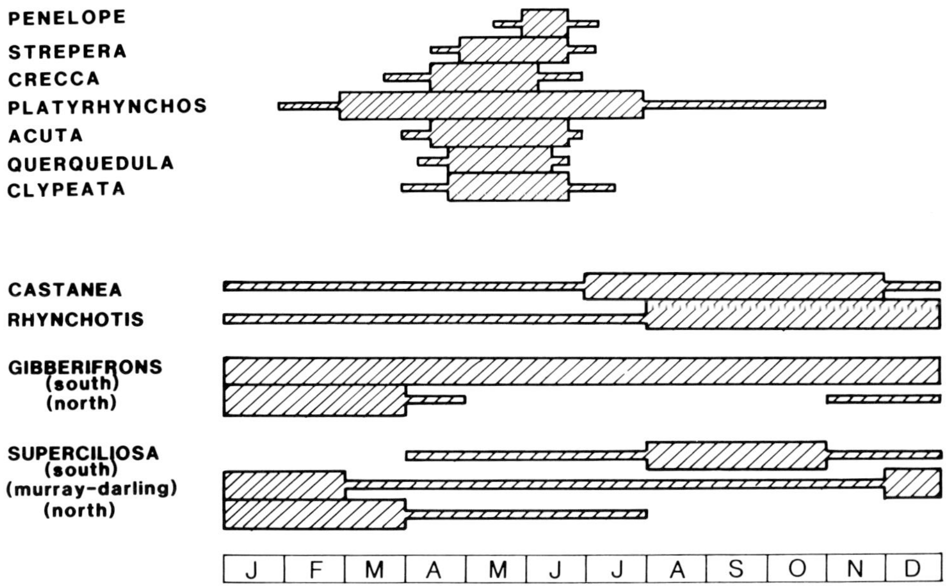

FIG. 3. Egg-laying seasons of seven holarctic and four Australasian *Anas* species (from Cramp and Simmons 1977; Braithwaite 1976a).

(*A. superciliosa*) tend to have regular annual breeding seasons, but Grey Teal breed at any time of year when conditions are favorable (Fig. 3). In all four species, breeding may be inhibited during periods of drought, and there are regional variations within species in the months when most breeding can be expected. The Grey Teal is an ecological counterpart of the Cape Teal in Africa, both being adapted to arid country through nomadism and irregular breeding. Braithwaite (1976a, b) has shown that the gonad cycle of Grey Teal is largely under environmental control (rather than photoperiodic and endogenous as in regular, annual breeders), and both sexes are capable of rapid gonad growth in response to rainfall, rising water levels, and the associated flush of aquatic food. If conditions remain favorable, some birds can raise two broods before molting. Therefore, in Grey Teal populations there is likely to be more marked asynchrony in the timing of breeding and molting than occurs in holarctic species with a short, regular, annual breeding season followed by a regular post-breeding molt (Fig. 3).

In addition to the wing-molt period, *Anas* females are "unavailable" for fertilization by males while they are incubating and brood-rearing, but females may be able to begin egg production again shortly after losing a clutch or brood. Therefore, at any one time, in a species with extended breeding seasons, a local population (e.g., of Grey Teal) might include some females that are fertilizable (pre-laying, laying or in the renest interval) and others that are not (incubating, brood-tending, post-breeding, molting). This could produce wide variation in operational sex-ratios (i.e., "the average ratio of fertilizable females to sexually active males at any given time": Emlen and Oring 1977) and males are likely to have more varied options than are open to males of holarctic species (Fig. 4).

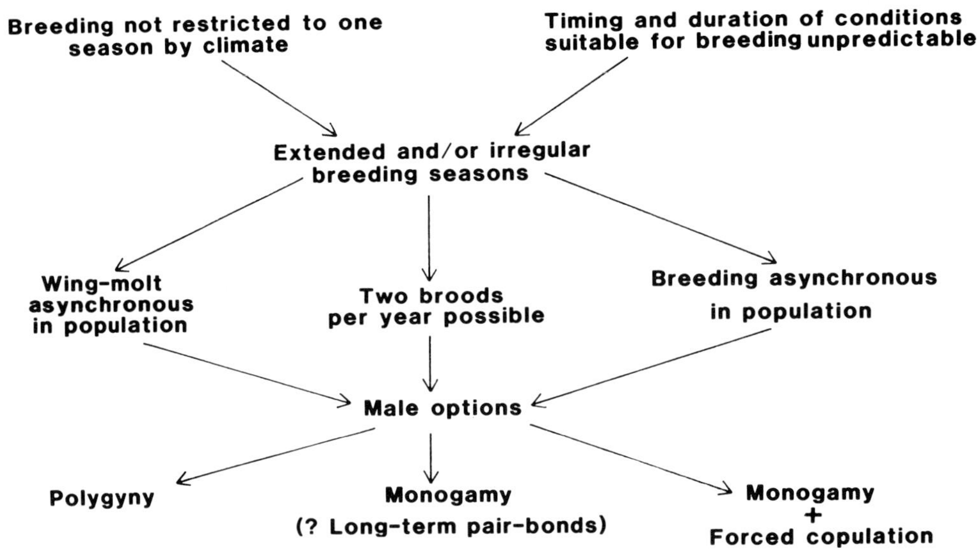

FIG. 4. Predicted male breeding options in southern hemisphere *Anas* species.

I believe that the bigamous tendencies shown by our captive Cape Teal, Speckled Teal, and White-cheeked Pintails reflect bet-hedging strategies in males of these species. For example, a male might be able to take a second mate while his first is incubating. If his primary mate succeeds in bringing off a brood, he may return to her and assist in brood-care. If the primary mate loses her clutch or brood and does not initiate another breeding attempt promptly, the male could switch to a second mate if the latter is close to egg-laying condition. Females are especially vulnerable to desertion while in wing-molt, but then so are males. African Black Ducks associate with partners other than their mates while the latter are molting and such liaisons could lead to mate-switches (McKinney et al. 1978).

Little is known about sex-ratios in southern hemisphere dabbling ducks. Data for adults and juveniles of African Yellowbill and Red-billed Teal in birds trapped for banding (Dean and Skead 1977) and for Grey Teal, Australasian Black Duck, and Australasian Shoveler from hunters' bags (Braithwaite and Norman 1974, 1976, 1977) are either skewed in favor of males (similar to ratios in northern hemisphere species) or show no significant difference. However, these findings may have little relevance to the operational sex-ratio. It is the availability of fertilizable females at different stages in the breeding season that is expected to influence the potential for polygyny.

The important influence that degree of synchrony among the individuals in a breeding population is likely to have on mating systems is now widely recognized (Maynard Smith 1977) but few studies have documented such effects (Wunderle 1984). As Wells (1977) has pointed out, in regard to anuran amphibia, it is essential to distinguish between the "breeding season" of a species, the "breeding period" of local populations, and the "breeding histories" of individuals. To understand the mating system, we need to focus on individuals and their reproductive efforts over their lifetimes. Intensive studies of marked birds are required to assess whether multiple mates are economically defendable (i.e., the "environmental potential for polygyny": Emlen and Oring 1977) and to determine the extent to which individual males can exploit this potential.

Comparison with Other Avian Mating Systems

Female-defense polygyny appears to be a relatively rare type of mating system in birds (Oring 1982). It is known in rheas, tinamous, and some pheasants, but in various ways all of these are different from what we have observed in *Anas*. Emlen and Oring (1977) expect female-defense polygyny to evolve when females clump for reasons other than reproduction (e.g., safety, localized resources) and the females in a harem are expected to cooperate. It is possible that an element of "resource-defense" is involved in some *Anas* species. For example, Speckled Teal are reported to use cavities in nests of the Monk Parakeet (*Myiopsitta monachus*) as nest-holes in eastern Argentina (Weller 1967), and this could produce clumping that would allow a male to maintain bonds with two or more females simultaneously. Cape Teal nest on the ground but may prefer nesting on islands (Winterbottom 1974), which could result in similar clumping of females. Nothing is known about patterns of nest dispersion in wild White-cheeked Pintails.

Among other waterfowl, polygynous behavior has been described in one population of the Comb Duck (*Sarkidiornis melanotos*), a member of the perching duck group (tribe Cairinini) with races in tropical Africa and South America. Siegfried (1979) observed males of the African race defending territories, advertising for mates from tree-top perches, and holding harems of up to four females. Females nest in tree-holes and Siegfried suggests that the distribution of suitable cavities is important in attracting females to certain territories. He observed males competing for females, and females competed with each other for cavities. Unmated males apparently clustered around harems and tried to court or make FC attempts on females in the harem.

Possibly other members of the perching duck group also have polygynous systems. All are hole-nesters, which could produce clumping of females, and most are tropical and have extended breeding seasons.

Forced copulation is apparently a widespread phenomenon, not only in *Anas* but also in other waterfowl tribes. In certain holarctic species it appears to be a secondary male insemination strategy but our observations on bigamy in three southern hemisphere species suggest that in addition to an insemination function, males might be using FC to establish relationships with females as a prelude to taking them as secondary mates. On the other hand, Siegfried's (1979) description of FC activities in Comb Ducks suggests that an "alternative mating strategy" (Dawkins 1980; Rubenstein 1980) might be involved here, subordinate unpaired males adopting a sneaking strategy. Studies of marked birds in the wild are needed to establish whether this is the case.

Conclusions

1. Monogamy is the primary mating system in holarctic dabbling ducks probably because each male is unable to monopolize more than one female in synchronously breeding populations with adult sex-ratios skewed in favor of males.

2. The environmental potential for polygyny (EPP) is expected to be relatively higher in many southern hemisphere *Anas* populations, which have extended breeding seasons and asynchrony in breeding and wing molt.
3. Bigamous tendencies observed in males of three southern hemisphere species in captivity suggest that males of these species can exploit this EPP.
4. The behavior of female Cape Teal and Speckled Teal involved in bigamous relationships indicate strong rivalry for monopolization of the male, and suggest a competitive female-defense polygyny system.
5. Forced copulation appears to be the only secondary reproductive strategy available to males of holarctic species; in southern hemisphere species FC can occur as well as polygyny and the two options may be intimately related.
6. Field studies on marked birds of southern hemisphere species are needed to test these ideas.

ACKNOWLEDGMENTS

D. Bruggers, J. T. Burns, K. M. Cheng, S. Evarts and P. Stolen collaborated in the flight pen studies on several species. For reading and commenting on the manuscript I am grateful to L. W. Braithwaite, D. Bruggers, P. Gowaty, D. Mock, F. I. Norman and M. W. Weller. My research has been supported by the National Science Foundation grants (GB-36651X, BNS 76-02233 and BNS-7924692), the National Geographic Society, and the Graduate School and Field Biology Program of the University of Minnesota.

CHAPTER 7

ADAPTIVE SIGNIFICANCE OF MONOGAMY IN THE TRUMPET MANUCODE *MANUCODIA KERAUDRENII* (AVES: PARADISAEIDAE)

BRUCE BEEHLER[1]

Biology Department, Princeton University, Princeton, NJ 08540

ABSTRACT.—The socioecology of the monogamous Trumpet Manucode (*Manucodia keraudrenii*) was compared with that of two polygynous species of birds of paradise, the Magnificent Bird of Paradise (*Diphyllodes magnificus*) and the Raggiana Bird of Paradise (*Paradisaea raggiana*) on a forest site in the mountains of eastern New Guinea. Information on diet, food dispersion, and breeding behavior were examined to test the applicability of various hypotheses pertaining to the ecology of mating systems in tropical forest birds (cf. Snow 1976; Lill 1976; Snow and Snow 1979; Emlen and Oring 1977; Bradbury 1981). Two points of focus in these theories are: male emancipation from nesting duties and the costs of polygynous court display.

The Manucode was found to be a fig specialist and was almost exclusively frugivorous. Even its nestlings were fed a diet of mostly figs and few or no arthropods. In contrast, the two polygynous birds of paradise took fewer figs, many specialized capsular varieties of fruit, and significantly more arthropods. In addition, they fed their offspring a mixed diet of arthropods and fruit. The ramifications of the dietary differences between the Manucode and the polygnous pair are two-fold: (1) the carbohydrate-rich but protein- and lipid-poor fig diet of the Manucode nestlings may promote biparental attendance; both parents may be needed to provide sufficient fig pulp for nestling growth. And (2) the rarity, asynchrony, and non-annual cycle of a number of fig species favored by the Manucode may produce foraging demands that make it uneconomical for males of this species to pursue the time-consuming and energetically costly routine of fixed-site polygynous courtship display used by the majority of paradisaeid species.

Available theory proves of minimal predictive power, at least when superficially applied to the birds of paradise treated here. In rainforest passerines, the importance of the nutritional quality of a frugivorous diet, and the coevolutionary relationship between food plant and avian seed disperser are two important influences in the evolution of male emancipation and polygynous mating systems that have not been accounted for in theoretical models.

INTRODUCTION

A fundamental question in avian sociobiology is the adaptive basis of interspecific variation in breeding organization. This variation may manifest itself in nesting habits, pair-bonding, or mating interaction. In this paper I focus on variation in avian mating systems and, specifically, I examine monogamy in the birds of paradise, an Australasian group of songbirds in which polygyny predominates. Monogamy is the dominant breeding organization in birds, and this is presumably because adaptive conditions strongly favor this breeding strategy, one that is nearly absent among mammals (Kleiman 1977; see Introduction, Chap. 1, for details).

[1] Present address: NHB Room 336, Smithsonian Institution, Washington, DC 20560.

Most recent studies of avian mating systems have focused on polygynous or polyandrous systems in an attempt to discover what conditions promote these atypical habits (Snow 1962; Zimmerman 1966; Orians 1969; Jenni and Collier 1972; Wiley 1973; Lill 1974, 1976; Oring and Maxson 1978). From these studies a broad array of hypotheses have been generated to explain the evolutionary significance of different mating behaviors (Orians 1969; Lill 1976; Snow 1976; Emlen and Oring 1977; Oring 1982).

The overwhelming consensus is that environmental conditions strongly influence the organization of mating systems. John Crook's studies of the evolution of social behavior in the weaverbirds (Ploceidae) were important because they provided convincing examples from an ecologically and behaviorally diverse bird family to support the argument that diet influences social behavior (Crook 1964, 1965). Numerous subsequent studies have broadened the empirical basis for this generalization. Recent reviews of the literature treating vertebrate mating systems have been valuable in cataloging and classifying the range of variation in monogamous and polygamous mating systems, and have offered several models to explain trends in the evolution of mating behavior (Emlen and Oring 1977; Wittenberger 1979; Bradbury 1981; Oring 1982). Studies of polygynous neotropical forest birds by Lill (1976) and Snow (1976) provide additional theoretical speculation on the basis of variation in mating organization.

Most of the major theories of mating systems have been summarized in preceding chapters of this monograph. In the following section I shall briefly touch on those most relevant to the present study. Most birds are monogamous because nesting productivity depends strongly on parental investment, and individual reproductive output is closely related to efforts of parental provisioning of offspring (Trivers 1972; Maynard Smith 1977). In situations where only one parent is needed at the nest, it is usually the female that assumes these duties. The female produces many fewer gametes than the male, but these few are larger, more richly provisioned, and expensive to produce. There is a fundamental sexual difference in the potential rewards of a copulation. The female benefits from being conservative and discriminating. In contrast, the male often can enhance his own reproductive output by mating polygamously.

In birds of paradise, as well as in many other bird families, considerable interspecific variation is found in parental investment by the male. The male either remains monogamous, helping at the nest, or else he gives up parental duties to pursue a strategy of attempting to attract and court multiple females. Thus the crucial behavioral shift is that of "male emancipation" from nesting duties. To understand the typical polygynous mating system, the mechanics of this shift in male parental investment must be explained.

Naturally, male emancipation can occur only under conditions where the solitary female can successfully provision her brood (Snow 1976; Lill 1976). This depends, in part, on being able to harvest sufficient food for herself and her nestlings. It is further postulated that the feasibility of such a scheme varies, depending on the diet of the bird: a single female foraging for elusive and difficult-to-harvest foods, such as arthropods (Lill 1976), will have difficulty successfully provisioning the nestlings, whereas frugivores or frugivore/insectivores will have greater ease in meeting the demands of single-parent nesting. This argument is based on the idea that fruit is abundant and relatively simple to harvest, compared with arthropod prey. Additionally, Snow (1976) postulated that frugivory allows less time to be spent on foraging and more time to be spent on other pursuits, such as mate acquisition, which further promotes the polygynous lifestyle.

Finally, Bradbury (1981) noted that dietary considerations produce patterns of spatial use that may influence the potential for polygamous mating by males. Wide-ranging females with high individual overlap of ranges raise contact rates between the sexes, whereas sedentary habits and exclusive territoriality reduce rates of contact, and lower the potential of promiscuous males encountering and mating with many females. Bradbury showed that different diets could promote different rates of intersexual contact, and thus alter the "polygamy potential."

In this paper, I will examine the validity of these ideas about frugivory, male emancipation, and mating systems in birds of paradise. In the final section I will offer some ideas about the applicability of generalizations derived from my study of birds of paradise on Mount Missim to theory on mating systems among other tropical species.

The study.—The birds of paradise (Paradisaeidae) are a family of 42 species of jay- to crow-sized forest birds of New Guinea, eastern Australia, and the Moluccas. Available data suggest that at least 31 of the species in this family are polygynous (Beehler and Pruett-Jones 1983). In such a family, where monogamous behavior is the exception to the general trend, it is instructive to examine what pecularities characterize those few species that stray from the paradiaeid "norm" of polygynous arena display. In this way I will focus on the nature of monogamy and attempt to test the ideas of Snow, Lill, Bradbury, and others concerning the environmental factors favoring this habit in an avian group where monogamy is a "peculiarity."

I examined the behavior and ecology of three species of birds of paradise: the monogamous Trumpet Manucode (*Manucodia keraudrenii*), and two polygynous species, the Magnificent Bird of Paradise (*Diphyllodes magnificus*) and the Raggiana Bird of Paradise (*Paradisaea raggiana*). The hypotheses of Snow and Lill predict that, all things being equal, the Manucode should face environmental constraints that demand biparental care at the nest, or else that prevent a male from attempting a polygynous strategy even if male emancipation is feasible. The presence of a male at the nest does not necessarily mean that uniparental care is impossible. It simply may indicate that, given the various alternatives, male help at the nest is the most productive option for his long-term inclusive fitness.

METHODS

From July 1978 through November 1980 I studied birds of paradise on a plot of forest at 1450 m altitude on the southwestern slopes of Mount Missim, Morobe Province, Papua New Guinea (7°16′S, 146°42′E). The plot was in original forest that is relatively aseasonal in rainfall and temperature (Beehler 1983a). The 17-ha plot straddled a ridge in well-drained, hilly, forested country with ridge-and-ravine topography. The birds were studied in an undisturbed state, although the local human inhabitants also use the forest for hunting the larger game birds and feral pigs. I was assisted in various aspects of my study by two, and occasionally three, indigenous field technicians, who provided essential skills related to handling mist-nets in the canopy, and finding bird nests and display sites.

In this study I focused on three aspects of the birds' life histories: (1) diet, (2) spatio-temporal dispersion of foods, and (3) breeding behavior. Below I summarize the field techniques.

Diet.—Most (and probably all) birds of paradise consume both fruit and arthropods, and, on occasion, small vertebrates such as frogs and lizards (Gilliard 1969; Schodde 1976). I used two methods for documenting interspecific differences in diet: (1) collection of fecal samples and (2) systematic monitoring of the birds foraging at fruiting plants.

To document the relative importance (by volume) of fruit and arthropods in the birds' diets, I assayed the contents of fecal samples collected from mist-netted birds. Individuals were placed in a darkened holding cage with a wire-mesh floor and excreta were collected and preserved in small vials of 70% ethanol. In the laboratory each labeled sample was washed and separated into its fruit and arthropod components. The volumes of these fractions were then measured by visual estimate by spreading the two components on a gridded petri dish and comparing number of covered squares. Nearly one half (27 of 56) of the samples were either 100% fruit or 100% arthropods.

To obtain detailed information on fruit-feeding preferences, I made repeated observations at fruiting trees to monitor the birds' foraging activities. In addition to the species visiting the fruiting plants, data were collected also on foraging rate, length of feeding bout, and interspecific aggression. Further description of my methods and a detailed analysis of these data appear elsewhere (Beehler 1983a, b).

Measuring food availability.—I used five methods to estimate the spatial and temporal distribution of food on the study area: (1) phenological survey of selected food plants, (2) fruit trapping, (3) mapping of all trees within two adjacent 1-ha plots, (4) seasonal surveys of fruit production by all species on the 17-ha study site, and (5) detailed mapping of populations of eight species of paradisaeid food plants on the study area.

Early in the project I ascertained the location and identity of fruiting plants favored by birds of paradise and selected 110 plants of 29 species for monthly phenology studies. Further data on seasonal cycles of fruiting were obtained from 50 fruit traps arrayed through the forest and checked on a biweekly schedule (see methods of Pratt 1983).

Spatial patchiness of food plants was measured on two scales. I mapped and identified all trees larger than 15 cm DBH within two 1-ha plots in the forest. In addition, every four months I carried out intensive seasonal mapping censuses of all plants producing fruit within the 17-ha study site. Each census required ca. 70 man-hours. Survey lines were walked through every section of the study site, and all plants with ripe fruit were recorded on a 1:3300 scale topographic map. Near the end of the study I made a special effort to map every individual of eight species of food plants that were most important to the birds of paradise. This mapping project was integrated with the fruit censuses. Fruiting periodicity was documented using data from the fruit traps, phenology trees, opportunistic reports by members of the field team, and from fecal samples. Information on spatial dispersion of the food plants came from the detailed plant mapping, intensive censusing, and opportunistic searching for particular species.

Insects compose an important part of the diet of many birds of paradise (Schodde

TABLE 1

ASPECTS OF SEXUAL DIMORPHISM, MATING ORGANIZATION, AND NESTING AMONG THE THREE BIRDS OF PARADISE[1]

Character	Bird of Paradise		
	Magnificent	Raggiana	Manucode
Sexual dimorphism	great	great	minimal
Mating interaction	polygynous	polygynous	monogamous
Display site	terrestrial court	communal lek tree	unfixed, arboreal
Male defends	display court	lek perch (= rank?)	(female?)
Parental care/nesting	female only[2]	female only	male and female
Vocalization	male only	male only	male and female

[1] Data from Mt. Missim studies, supplemented by published information, where noted.
[2] Everitt 1965; Bulmer, in Gilliard 1969.

1976). This portion of these birds' diets is considered in greater detail elsewhere (Beehler 1983a). For the purposes of these analyses, the proportion of arthropods in the diet is considered as the key measure of insectivory.

Courtship behavior and nesting. —Although descriptions of courtship behavior have already been published for these three species (Gilliard 1969), I gathered an additional 201 h of behavioral observations on the Magnificent Bird of Paradise and 71 h of observations on the Raggiana Bird of Paradise. Because the Manucode's displays and courtship are brief and unpredictable, I was unable to make systematic observations, but I observed Manucode mating behavior opportunistically. With the aid of field assistants I searched for active nests. When found, a nest was studied from a blind on the ground, observing with a 15× telescope. I monitored parental attendance (e.g., sex roles) and food provided to nestlings, with special efforts to obtain information on the types of foods nestlings consumed. The nesting Manucodes had been color-banded and identified to sex by Thane K. Pratt, who was carrying out another study on Mt. Missim.

RESULTS

Comparisons of behavior. —The two polygynous species share several morphological and behavioral features that differ from those of the monogamous Manucode (Table 1). Sexual dimorphism is evident in both polygynous species, with males showing elaborate nuptial plumage used in courtship display. The sexes of the Manucode are superficially alike and are relatively unornamented.

A fixed and traditional site of display and mating makes it simpler for females to find and evaluate potential mates (Bradbury 1981). Display sites are fixed for the two polygynous species, but not for the Manucode. Males of the polygynous species attend their sites every day during the display season, which lasts more than six months per year. The male Manucode has no fixed or traditional site for courtship; copulation occurs after a comparatively brief and simple nuptial display that can take place at any locale within the home range (Hoogerwerf, in Cooper and Forshaw 1977; Beehler 1983a).

Resource defense among the three species provides no obvious contrasts. Unlike many temperate insectivores (e.g., Song Sparrow [*Melospiza melodia*] and others; see Oring 1982, for examples), these birds of paradise defend no foraging or nesting

TABLE 2

RELATIVE ABUNDANCE OF ARTHROPODS IN FECAL SAMPLES OF THREE BIRDS OF PARADISE[1]

Species	Number of samples	Frequency distribution of samples with different proportions of arthropod material							Mean[4] for all N
		All arthr.	0.9–0.7	0.6–0.5	0.4–0.3	0.2–0.1	0.1	All fruit	
Magnificent[2]	31	2	2	5	6	2	3	11	0.3 (±0.3 s.d.)
Raggiana	12	—	—	1	1	4	3	3	0.1 (±0.2 s.d.)
Manucode	9[3]	—	—	—	—	—	2	7	0.01 (±0.02 s.d.)

[1] Here measured as a proportion of each sample, by volume. See text for further explanation.

[2] Interspecific comparisons were made between the samples. A Mann-Whitney *U*-test, one-tailed, was used to compare the Magnificent to the Raggiana: $U = 253.5$, $P > 0.03$, N.S., comparing the combined data for the Magnificent and Raggiana against the Manucode: $U = 308.1$, $P < 0.001$. Assigned level of significance = $P = 0.01$ (Siegel 1956).

[3] Data for three Manucode samples from Schodde (1976).

[4] These means are calculated from the actual measurements taken, which in this table have been classified into seven categories for convenience of tabulation. Accuracy of measurement increased as the proportion of arthropod matter in the sample decreased below 0.2. Thus data for the Manucode are noted to the level of 0.01.

territory. All three species are primarily frugivorous (see below), and it is apparent that competition for mates is not based on resource control (sensu Emlen and Oring 1977).

The Manucode exhibits biparental care at the nest with both the male and female attending and feeding the nestlings. Available evidence shows that individual pairs of Manucodes may remain together for more than a single season (T. K. Pratt, pers. comm.). In the two polygynous birds of paradise, the female constructs the nest and rears the nestlings unaided. That intersexual cooperation is more important in the monogamous species is further indicated by data on vocalizations. For the two polygynous species the male is highly vocal while the female is almost mute. The male vocalizes to advertise his presence in the forest and, presumably, to attract potential mates. In contrast, both male and female Manucode deliver a variety of sex-specific vocalizations. Pairs occasionally perform rudimentary duets.

Comparison of diets.—Examination of fecal samples showed that the two polygynous species took significantly more arthropods (by volume) than the Manucode (Table 2). Although none of the three took more arthropods than fruit, this insect component averaged 30% of the Magnificent, 10% of the Raggiana, and 1% for the Manucode.

I recorded 836 feeding bouts by the three paradisaeid species at 29 species of plants, from 14 botanical families (details in Beehler 1983b: Table 2). Morphologically, the fruits show considerable interspecific variation. For comparison, I classified fruit of the 29 species into three groups: capsular species, drupe/berry types, and figs (Beehler 1983b). The first two categories are non-taxonomic and primarily morphological. Capsular species have the edible arillate "reward" encased in a dehiscent inedible protective capsule. In the drupe/berry fruits the edible portion surrounds the seed; this type is eaten whole and lacks any protective husk. The fig category includes only species of the genus *Ficus,* which are structurally similar, but nonetheless vary in size and color (Janzen 1979; Wiebes 1979). Of the 29 species of fruit taken by the three birds of paradise in this study, 10 species were capsular (N = 288 foraging observations), 10 were drupe/berry (N = 163), and 9 species were figs (N = 385).

TABLE 3

FRUIT PRODUCTION AND DENSITY OF FOOD PLANTS IMPORTANT TO THE BIRDS OF PARADISE (BOPS)[1]

Food plant	Plants/ha	Crop size per tree	Leng. fruiting season (mo)	% taken by BOPs	Fruit availability/day[3,4]
Ficus gul	0.12	50,000[2]	2	85%	700
Ficus #275	0.12	100,000	1.5	35	775
Homalanthus	1.2	12,000	5	93	75
Gastonia	0.12	40,000	1.5	100	900
Chisocheton	0.29	4,500	3	100	50
Endospermum	0.47	30,000	2.5	57	225
Elmerrillia	0.59	15,000	1.5	40	125
Dysoxylum	0.35	2,500	3	100	25
Ficus odoardi	0.65	5,000	2.5	95	65
Cissus	0.3±	15,000	2	54	135

[1] This list of 10 species is made up of the food plants most frequently visited by the three primarily frugivorous birds of paradise. The list includes the five most frequently visited plants for each of the three bird species. There is some overlap in choice. See Beehler (1983b) for complete taxonomic nomenclature for these plants.

[2] This figure is estimated from crops of trees encountered on the study area. In the last season of the study, a typical mature individual for each species was used as the sample. Fruit here is equivalent to edible portions. In most cases this equals a whole fruit or fruitlet, but in *Ficus odoardi,* which is eaten by breaking it into pieces, each large fruit is considered the equivalent of five edible portions.

[3] Calculated: total fruit crop/length of season × the percent taken by BOPs.

[4] For the favorite Manucode figs, given a measured population density of 1 plant/8 ha (Fig. 2), a fruit production season of 2 mo, total population asynchrony, and a non-annual fruiting cycle of 9 mo, the following calculation can be made. Each plant produces fruit for 2.7 mo/yr, on average. Under perfect conditions of no inter-plant overlap, 4.4 plants could provide fruit during the entire 12 mo of the year. Assuming, however, that chance permits 50% overlap in timing of fruit production, we double the number of plants needed to provide fruit during all months of the year, to 8.8 plants. The area needed to cover these 8.8 plants would be 70 ha.

The two polygynous birds of paradise took greater proportions of capsular fruit than any other category (Fig. 1). Capsular fruit made up 49% of the Raggiana's fruit diet (44% of total diet), 58% of the Magnificent's fruit diet (41% of total), but only 8% of the Manucode's fruit diet (8% of total). Figs, on the other hand, were rarely taken by the Magnificent (9% of fruit diet, 6% of total), moderately popular with the Raggiana (35% of fruit, 32% of total), and the dominant fruit source for the Manucode (80% of both fruit and total diets). Drupes represented small-to-moderate fruit sources for all three species.

In summary, the Manucode is a fig specialist and nearly an absolute frugivore. The two polygamous birds of paradise preferentially take capsular fruits and small-to-moderate quantities of arthropods.

Spatial dispersion of fruit resources.—Food plants of birds of paradise are relatively rare in the habitat. On the Mount Missim study plot, the oak *Castanopsis acuminatissima* (not a paradisaeid food plant) was present in densities as high as 50 plants per ha. In contrast, most paradisaeid food plants were present in low densities (fewer than one mature plant per ha; Table 3). Among paradisaeid food plants, however, there are differing degrees of rarity, from a low of 0.12 plants per ha for *Ficus gul* to a high of 1.2 plants per ha for the small euphorbiaceous tree, *Homalanthus novoguineensis,* which is irregularly scattered through the forest (Fig. 2) perhaps because it is a light-gap specialist. *Endospermum medullosum* is much less clumped. Thus, a range of spatial distributions among the food plants is evident (Fig. 2).

Temporal dispersion of fruit.—Timing and synchrony of fruiting are important factors that, combined with the spatial component, produce the resource cycle to which frugivores adjust seasonally. Three important determinants of fruit's tem-

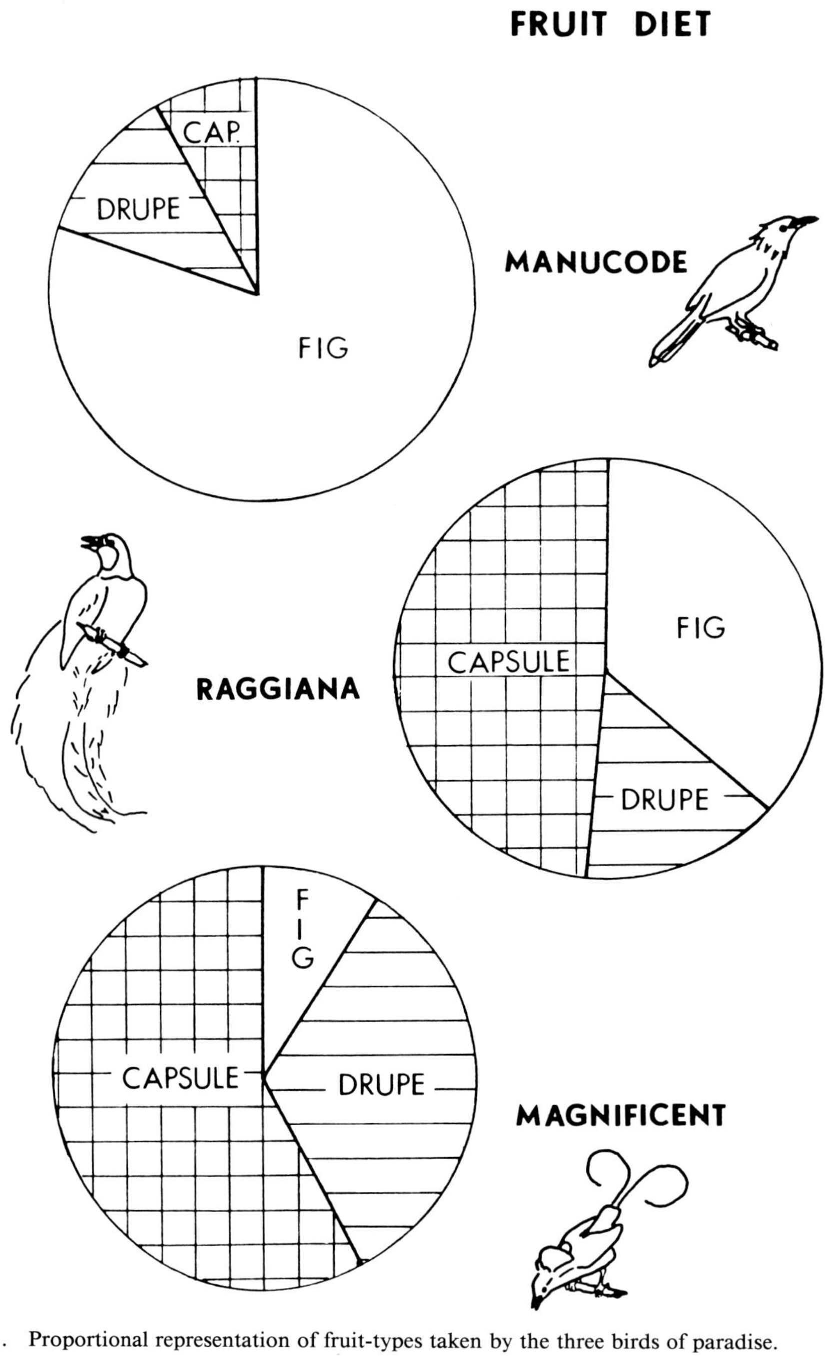

FIG. 1. Proportional representation of fruit-types taken by the three birds of paradise.

poral availability are: (1) length of each plant's fruiting season, (2) seasonal timing of its fruit production, and (3) synchrony of fruiting in relation to other plants. Fruiting seasons for individual trees are as short as 1.5 months for strangler *Ficus* species #275 to six months for *Homalanthus*. In most instances, length of

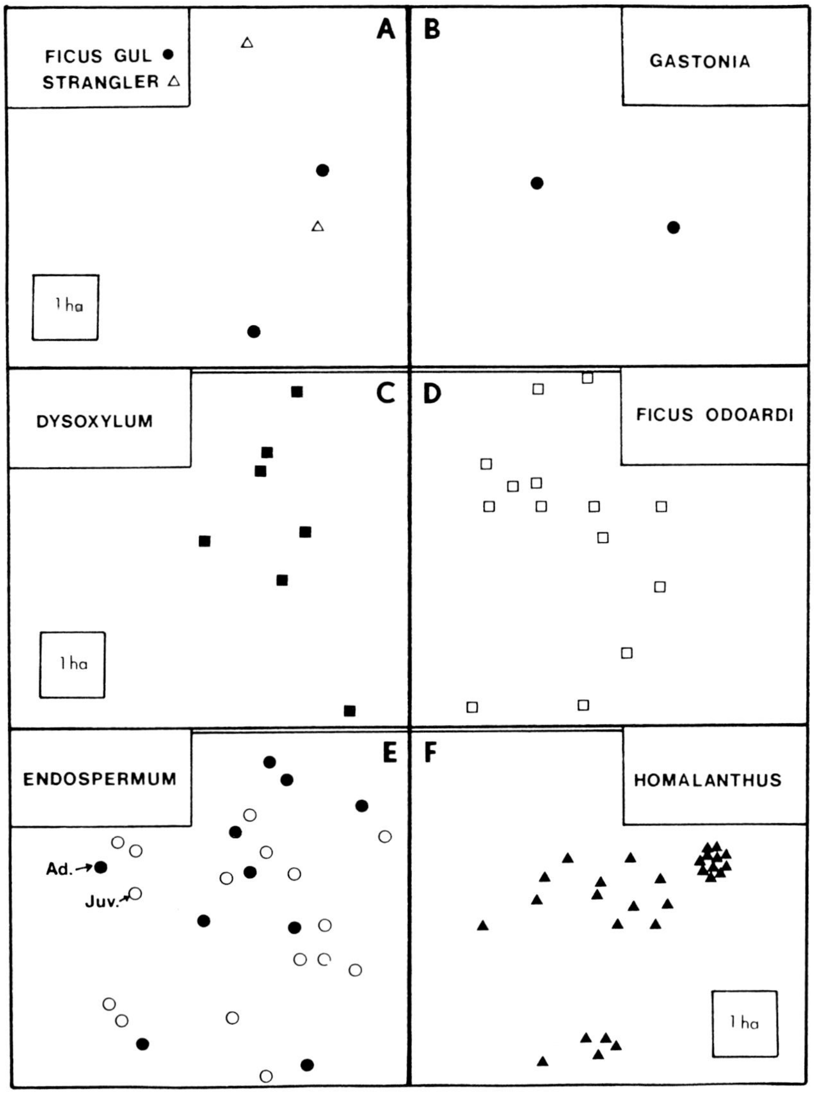

FIG. 2. Maps of the spatial distribution of seven species of food plants on the Mount Missim study plot used by the birds of paradise. Note scale from the one-ha quadrat.

fruiting season and percentage of the fruit crop available at any one point are inversely related. Thus, other things being equal, food plants with shorter seasons offer foragers more fruit per day, serving as richer "patches."

In contrast with other types of food plants, many fig species produce fruit throughout the year. Individual trees of a number of fig species produce crops on a subannual schedule (i.e., in some calendar years produce two crops) and most or all species show population asynchrony. The species *Ficus gul* and *F. drupacea* exemplify this pattern, showing no annual periodicity, and thus making an unreliable foraging resource. Because of extreme within-species asynchrony, one can find individual fig trees fruiting in any month of the year.

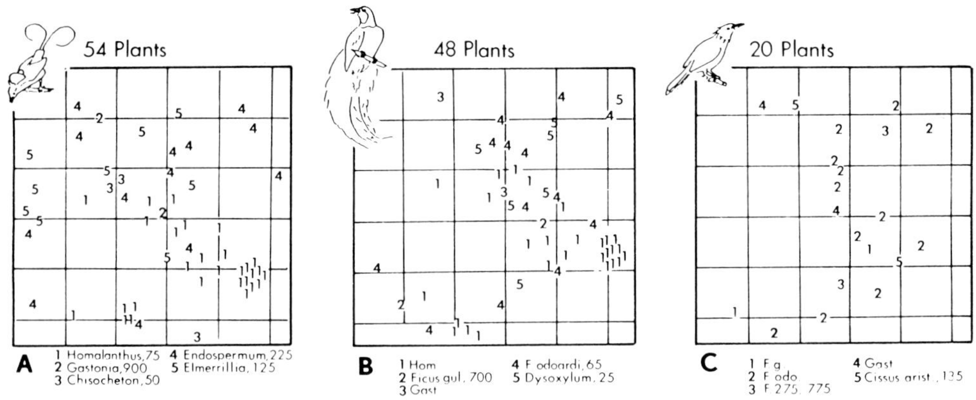

FIG. 3. Composite maps of fruit resource distribution of the three birds of paradise. The populations of the five most used food plants for each bird are mapped. Numbers accompanying the species-keys indicate estimates of daily fruit production (Table 3). A. Magnificent; B. Raggiana; C. Manucode.

By contrast to the figs, species like *Homalanthus, Pandanus conoideus,* and *Dysoxylum* produce reliably in the same month every year. All individuals of each local population fruit in the same season (high synchrony), making the fruit of each species available in relative abundance at one predictable season, then entirely absent throughout the rest of the year. Table 3 summarizes information on the five most important food plants for each of the three birds of paradise. These plant species vary considerably in crop size, length of fruiting season, and use by birds of paradise. To provide a static view of differences in the patchiness of fruit resources for each bird of paradise, the distribution of favored food plants is mapped in Figure 3. Within this ca. 25-ha map area, the two polygynous species could find 49 and 54 individuals of preferred food plants, whereas the Manucode had only 20 favored plants in the same area. Because the typical home range for each of the three species is as large or larger than 17 ha (Beehler 1983a), the size of the resource patch sampled reflects a realistic subset of the food plants used by the birds.

Information on the fruit crops (Table 3) of these favored food plants provides an added contrast in dispersion of resources for polygynous vs monogamous species. The five favored food plants of the Manucode show an average daily production of 525 fruits (for method of calculation see notes with Table 3 in Beehler 1983a). For the two polygynous species, the average crop production is 314 fruits per day. Because the average size of the edible portion of favored Manucode fruits does not differ significantly from that of the polygnous species (Beehler 1983b), this higher per-day production of Manucode fruiting plants reflects greater concentrations of food in fewer, more widely spaced patches.

DISCUSSION

The hypotheses of Snow (1976), Lill (1976), Emlen and Oring (1977), and Ricklefs (1980) predict that interspecific differences in mating systems should correlate with parallel differences in diet. They argue that dietary differences promote different strategies for nesting and mate acquisition. One diet might favor easy provisioning at the nest by a single parent, whereas another requires both

TABLE 4

CHARACTERISTICS OF TWO FRUITING PLANT STRATEGIES

Character	Fruiting type	
	Trapline	Opportunist
Nutrition	rich in lipid and protein	rich in water and carbohydrate
Crop size	small to moderate	large to very large
Mean distance between patches	moderate	very high
Availability through year	moderate to long season; absent for part of year	short patch-life but available in different patches through year
Year-to-year patch predictability	high	low/absent
Overall annual production per plant	small to moderate	large to huge
Rate of renewal (per plant)	annual	subannual or annual
Examples from this study	*Chisocheton* *Dysoxylum* *Homalanthus*	*Ficus gul* *Ficus drupacea* *Ficus obliqua*
Used by what foragers?	mostly BOPs	many types

parents to work full time. In addition, one diet might be compatible with daily foraging from a fixed site (important for a bird with fixed-court display), whereas another diet might require heavy weekly ranging movements to locate rare food resources. To test the importance of the role of diet in paradisaeid mating behavior, I examined the relationships among diet, male emancipation, and mating organization for the three species studied.

Diet and behavior.—The Manucode and Magnificent have an average clutch-size of 2, the Raggiana 1 (Gilliard 1969; Cooper and Forshaw 1977). Nesting conditions appear similar for the three species. All produce an open cup nest placed on the outer branches of a tree (Gilliard 1969). No obvious interspecific differences related to nesting habits are known except that one species shows biparental care, whereas the other two show female-only care at the nest.

Two factors might be important in the role of male emancipation: (1) distribution of food for adults, and (2) distribution and nutrition of food for nestlings (Snow 1976; Snow and Snow 1979; Foster 1978). One of the clearest findings in the comparisons of paradisaeid diets was that the Manucode preferentially takes figs, whereas the two polygynous species preferentially take fruit of capsular species. The characteristics of these two fruit resources differ considerably, so they merit further discussion.

Opportunist vs trapline fruit.—A well-known dichotomy in fruiting strategies exists for food plants of vertebrates in the tropics (McKey 1975; Howe and Estabrook 1977; Bradbury 1981; Janson 1983). One strategy is to produce copious amounts of relatively "cheap" fruit for a wide spectrum of generalized foragers. Termed the "opportunist" strategy, this fruiting mode in the New Guinea montane forest refers to plants that attract foragers by producing large quantities of small-sized, small-seeded, and unprotected fruits which can be harvested easily by a wide range of birds from numerous families. The second strategy, termed "trapline," refers to the type of plants that serve a more restricted subset of presumably superior dispersers by producing smaller quantities of larger-sized, larger-seeded, protected (or morphologically specialized) fruits, whose crops ripen slowly over a long season (Bradbury 1981; Table 4). The trapline species produce more elaborate fruits, with better provisioned seeds and more nutritious disperser rewards. Only a few species of birds that can manipulate these fruits can utilize these specialized food plants (Howe and Smallwood 1982).

The term "opportunist" means that these food plants apparently depend for seed dispersal on the opportunistic use of fruit by a diverse array of foragers. "Trapline" is used because the plants of this type depend on a select segment of the avifauna to harvest the small quantity of fruit that ripens daily. To obtain a satisfactory daily supply of fruit, these foragers must visit a series of different plants, taking small samples from each. These two fruiting strategies produce food resources of distinctly different spatio-temporal availability and nutritional quality. Birds that feed preferentially on one or the other of these resource types can be expected to face different foraging challenges, which may influence social habits (Ricklefs 1980; Snow 1980).

As with most biological phenomena, this dichtomy in fruiting types is not absolute. Many plant species exhibit characteristics of intermediate nature or share characters drawn from both extremes. Just as no bird of paradise is restricted solely to one fruit type or another, few plant species possess all the traits of one fruiting strategy or the other. Nonetheless, the phenomenon is real and many birds show real preferences. Of the 29 food plant species treated here (accounting for 78% of tree-visits taken during foraging observations), 16 can be readily classified either as primarily "opportunist" (6 spp.) or primarily trapline (10 spp.). About 74% of the Manucode's fruit diet was composed of "opportunist" species, whereas 76% of the two polygynous species' diets consisted of "trapline" species. In montane New Guinea these two categories are filled by specific subsets of the flora. The "opportunist" plants, in most cases, are figs, and the "trapline" species are largely capsular varieties.

A typical "trapline" fruit is relatively rich in lipid and protein, whereas "opportunist" fruit is rich in water and carbohydrates (McKey 1975; Crome 1975; Howe and Estabrook 1977; Foster 1978). "Trapline" resources tend to be distributed in larger numbers of small-sized food patches and are produced over a longer season, whereas the "opportunist" food plants tend to be rarer, but very large and ephemeral crops. Foragers use these two resources in strikingly different ways. Birds visit a "trapline" plant in small numbers, checking for ripe fruit on a daily basis. At an "opportunist" fruiting plant, flocks of foragers swarm over the tree during its short fruiting season, and depart rapidly after the crop is depleted. As predicted by Snow (1976) and Ricklefs (1980), the polygynous paradisaeid species prefer the specialized "trapline" fruit, whereas the monogamous species prefers the "opportunist" fruit. Two characteristics, nutrition and predictability, stand out as possible factors in this relationship.

Nutrition and nesting. — The Manucode prefers "opportunist" food plants, generally figs more than others. Most figs are nutritionally poor, especially in comparison with the specialized "trapline" fruits. For example, many of the arillate species have large seeds, lipid- and protein-rich arillodes, and specialized sets of avian dispersers (Crome 1975; Frost 1980; Foster and McDiarmid 1983). Given that the Manucode relies on figs and takes virtually no arthropods, it may face difficulties in obtaining sufficiently balanced nutrients for self-maintenance, breeding, and critical periods of nestling growth. Foster (1978) has shown that total frugivory is a viable alternative for an adult bird, but potentially limiting for nestlings because of problems associated with processing large quantities of fruit pulp, to extract the relatively rare but essential proteins and lipids.

It may be impossible for a single parent Manucode to process and feed its two offspring enough fig pulp for rapid growth to fledging. My observations at one nest showed that the male and female took turns bringing food and brooding nestlings. When the foraging parent arrived at the nest and regurgitated its fig pulp, its brooding mate departed. This pattern continued throughout the day. Although I observed 70 nest exchanges, only once did a parent return without providing some food to the nestlings. Analysis of fecal contents below the nest showed that more than 90% of the material was fig pulp. No parent fed any identifiable insect matter to the nestlings. One can speculate that for the Manucode, a strict frugivore that specializes on figs, two parents are needed at the nest because of the demands of providing their offspring with sufficient fig pulp to fledge quickly.

In the Magnificent and Raggiana birds of paradise, only the female tends the nest and the offspring are fed a diet of fruit and arthropods (Dharmakumarsinhji 1943; Everitt 1965; Beehler, unpub. data). Nestlings of the Raggiana and Magnificent are fed a more balanced diet that includes sizable proportions of arthropods and a variety of fruits. The single tending parent (female) can subsist on the most easily obtainable fruit at the time of nesting, but the nestlings are fed a mixed diet of arthropods which are considered difficult to harvest (Lill 1976), and fruit (which varies in nutrition and ease of harvest).

Suggestive parallels exist in other ornithological studies. For example, Donaghey (1981) provides critical data on parental effort and nestling growth for two Australian bowerbirds, the polygynous Satin Bowerbird (*Ptilonorhynchus violaceus*) and the monogamous Green Catbird (*Ailuroedus crassirostris*). Although both sympatric species have a clutch-size of 2, the pair of catbird parents feed their nestlings a diet that is two-thirds fruit (mostly figs), whereas the solitary female Satin Bowerbird parent provides a nestling diet that is mostly arthropods. Feeding rate for nestlings of the Catbird is 1.6 times higher than for the Satin Bowerbird, yet the young of both species fledge after 21 days (growth rates are unavailable for the Satin; Donaghey 1981). On the other hand, Snow (1973) showed that the polygynous White Bellbird (*Procnias averano*) has uniparental care and a nestling diet of fruit only. In this exceptional circumstance, the fruits fed to the nestling bellbird are highly nutritious, specialized varieties which Snow's analysis showed to be rich in lipid and protein.

Fig-eating and male courtship demands.—The peculiarities of avian specialization on figs may promote monogamy through its effect on nesting ecology, or it may affect the male's ability to establish a functional polygynous courtship system as exemplified by the other tropical forest lek-displaying species.

Fig plants are unusual for their remarkable coevolution with their Agaonid wasp pollinators. Janzen (1979) contends that the dynamics of this mutualistic relationship are responsible for the asynchrony of fruiting and flowering within populations. Efficient wasp pollination also may make it possible for some fig species to maintain populations in the forest at very low densities. At any given time, one will find individual trees displaying different stages in floral reproduction. This occurs because newly emerging female fig wasps leave a fig plant at one stage in this cycle (at which the figs are ripe) and then must immediately find suitable sites for ovipositing (at which the fig fruits are at the young, receptive stage). Details of the wasp-plant relationship are given by Wiebes (1979). The rarity of some fig species in the forest is almost certainly because of the inability of their tiny seeds to compete with the better provisioned (larger) seeds of the average "true forest" species, such as typical trapline fruit (Janzen 1979). The annual seed crop from a single fig plant of a species favored by the Manucode includes many millions of seeds. But mature individuals of the species are exceedingly rare. For example, the density of *Ficus gul* on the study area was approximately one per eight ha. The presence of a mature fig in the forest is a result of the statistically rare event where a disperser deposits a fertile seed in a situation where light, nutrients, and the lack of competitors for sunlight all coincide (Janzen 1979).

The intra-populational asynchrony of many fig species is important to avian foragers. Primarily, this asynchrony enables a forager to specialize on figs year-round (Fig. 4). By searching an area of about 50 ha, a Manucode can expect to find one ripe *Ficus gul* tree every month of the year (see Table 3, footnote 4). A rough estimate of the size of a home range of a male Manucode is 200 ha (Beehler 1983a), enough forest to support a population of *Ficus gul* with at least four ripe fruit crops being produced in any particular month of the year.

The effects of fig asynchrony contrast sharply with the annual populational synchrony characterizing typical "trapline" food-plants favored by the polygynous birds of paradise (Fig. 4), whose fruit is abundant through the forest only during a predictable annual season. Thus, in one six-month season *Homalanthus* fruit are abundant and widespread, but then are absent for the ensuing six months.

Because of the seasonal nature of typical "trapline" species, fruit specialization by other birds of paradise differs considerably from that of the Manucode. Whereas a Manucode finds figs throughout the year, a Magnificent Bird of Paradise exhibits seasonally predictable specialization, taking a small but specific suite of food-plant species, and shifting its choice of fruit from one season to the next as dictated by the phenology of the various food plants. Thus on Mt. Missim, the Magnificent concentrated on *Homalanthus* from July to September, on *Gastonia* from October to November, and on *Chisocheton* from December to January.

For the male Manucode, the rarity of fig plants in the forest (Fig. 2), in conjunction with their temporal unpredictability, produces a resource distribution that presumably requires special foraging efforts. With population asynchrony and non-annual fruiting of individuals, figs are locally concentrated but seasonally "elusive" foods. The forager never has the opportunity to develop an annual foraging schedule whereby food-plants are available at predictable sites during the same seasons, year after year. Conversely, feeding on the predictable "trapline" fruit, the Magnificent Bird of Paradise can rely on a predictable harvesting regime, which it presumably learns during its first few years of life.

The nature of the fig resource may help explain why the Manucode is wide-ranging, non-territorial, gregarious, and lacking in a traditional and fixed site of courtship display. The ability of the Magnificent and Raggiana birds of paradise to occupy traditional fixed sites of display for more than one half of each year depends on their ability to invest great amounts of time and effort in attendance, maintenance, and courtship activities at the traditional site. This elaborate courtship habit may be made possible by the predictable nature of their nutritious

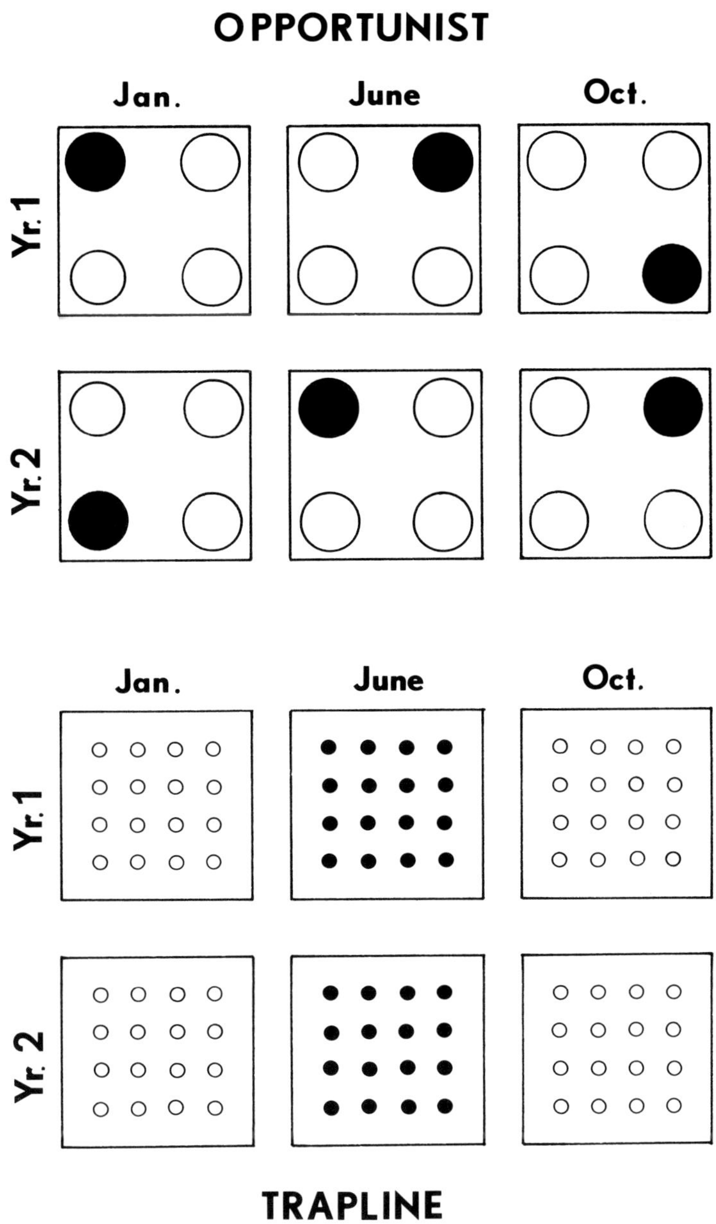

FIG. 4. Schematic representation of the patterns of fruiting of the typical "opportunist" food plant vs the typical "trapline" fruit. Open circles represent food plants without ripe fruit; solid circles = plants with ripe fruit. For this study, the typical opportunist plant was a fig; the typical trapline fruit was a capsular species from the family Meliaceae.

"trapline" fruit resource. In contrast, the male Manucode must renew the search for ripe figs, rare and unpredictable, each new season. The timing and distribution of ripe fig fruit changes every year. The demands of hunting for these elusive fig resources may not be concordant with the costly investments of polygynous lek display.

Monogamy in other birds of paradise. —The few monogamous birds of paradise for which there is information appear also to be specialized frugivores. The Crinkle-collared Manucode (*Manucodia chalybatus*), which shares the Mt. Missim study site with the Trumpet Manucode, is similarly a fig specialist and is monogamous (Gilliard 1969; Beehler 1983b). One other monogamous species, the crow-like Macgregor's Bird of Paradise (*Macgregoria pulchra*) is perhaps exclusively frugivorous; apparently it specializes on a single species of food plant, the gymnospermous tree *Dacrycarpus compactus* (Beehler 1983c), and seems to be restricted to the high altitude subalpine habitats where this tree grows.

Birds of paradise and mating system theory. —The Trumpet Manucode is by no means the prototypical monogamous species. Ostensibly, its wide-ranging, frugivorous habits would make it favorable for males to attempt a polygynous lek-displaying strategy, as exhibited by other polygynous tropical forest frugivores (Snow 1976; Emlen and Oring 1977); but one can understand the causal relationship between the bird's diet and its monogamous habits after looking closely at the ecological conditions under which the Manucode lives.

Used in a superficial way, dietary generalizations like "frugivore" and "insectivore" can cover a multitude of ecological conditions. The frugivory of the Manucode involves nutritional and energetic costs totally unlike those for frugivory in the Raggiana or Magnificent birds of paradise. The costs and benefits of insectivory in a primarily frugivorous species (like the Raggiana) are entirely different than those for an entirely insectivorous species. To the primary frugivore, insect-eating provides a back-up source of key nutrients needed by offspring; to the strict insectivore, it symbolizes the long hours of foraging each day to obtain needed energy as well as limiting nutrients. The "specialized frugivory" of the Manucode is similar to frugivory by most parrots and pigeons. All three avian examples rely primarily on fruit, but not in a fashion that might promote beneficial coadaptation between food plant and forager. Most parrots and pigeons are seed predators. The Manucode specializes on a food plant that is simply not evolutionarily tuned to specialized seed dispersal. Thus neither Manucode nor the parrot/pigeon assemblage gain from benefits that might accrue through a plant-bird mutualism, the special relationship in which many of the polygynous birds of paradise appear to belong (Beehler 1983b).

For birds of paradise, the diet that apparently promotes polygynous lek behavior is a mix of insects and fruit, the latter consisting of largely "trapline" varieties, but usually supplemented by "opportunist" varieties as well (Beehler 1983b). The "opportunist" fruits provide the abundant source of cheap energy. The "trapline" fruits provide a food source that is spatially and temporally predictable, as well as nutritionally rich. The arthropods provide valuable nutrients for growing nestlings. These different food resources combine to provide an apparently ideal foraging environment for the evolution of male emancipation of polygynous lek behavior.

For the Manucode, the demands of provisioning nestlings and the costs of male courtship display appear to be adaptive barriers that make monogamy the economical pathway for both male and female reproductive fitness. In this respect, monogamy in the Manucode is probably similar to that phenomenon in many other monogamous bird species. Although monogamy is the exception among the birds of paradise, it remains the phenomenon whose existence is simpler to explain in terms of its adaptive function in avian species.

For a full understanding of the evolution of mating systems in tropical forest passerines, more study is needed on the nutritional and energetic demands of provisioning altricial nestlings, and the relationship between foraging ecology and the demands of male attendance of a polygynous lekking site.

ACKNOWLEDGMENTS

For help with fieldwork in Papua New Guinea I thank Ninga Kawa, Michael Yamapao, the staff of the Wau Ecology Institute, and the Wildlife Division, Office of Environment. Thane K. Pratt, J. R. Croft, E. E. Henty, and D. G. Frodin provided the botanical identifications. This work constitutes a portion of a Ph.D. dissertation submitted to Princeton University, and I thank J. W. Terborgh, H. S. Horn, R. M. May, S. D. Ripley, and D. I. Rubenstein for guidance and advice. The project was funded by the Smithsonian Institution, Princeton University, National Geographic Society, and Frank M. Chapman Fund of the American Museum of Natural History. M. S. Foster, R. F. Pasquier, D. W. Mock, L. W. Oring, P. A. Gowaty, and an anonymous reviewer provided numerous and helpful criticisms on earlier drafts.

CHAPTER 8

THE INFLUENCE OF DEMOGRAPHY ON THE EVOLUTION OF MONOGAMY

BERTRAM G. MURRAY, JR.
Department of Biological Sciences, Rutgers University, New Brunswick, NJ 08903

ABSTRACT.—According to my theory on the evolution of mating systems (Murray 1984), the kinds of mating relationships occurring within a population are constrained by the individual's probabilities of surviving to later breeding seasons, the female's potential annual fecundity, and the adult sex-ratio. Females are often confronted with a choice between (a) mating with an already mated male because unmated males on suitable territories are not available and (b) waiting until a later breeding season to breed with a monogamous male. The cost of the former is reduced paternal care in rearing the young, but the cost of the latter is a greater average annual fecundity. I propose that a female accepts mating with an already mated male when suitable unmated males are unavailable and she cannot afford to wait until a later breeding season because her minimal annual fecundity is greater than her potential fecundity. This theory is illustrated by comparing the behavior of Prairie Warblers (*Dendroica discolor*) and Laysan Albatrosses (*Diomedea immutabilis*).

INTRODUCTION

The demography of populations has been all but ignored in developing theories accounting for the evolution of mating systems. A few early authors did suggest that unbalanced sex-ratios were responsible for polygamous mating relationships (Skutch 1935, 1976; Armstrong 1947; Williams 1952), and Maynard Smith and Ridpath (1972) thought that an unbalanced sex-ratio must have been the initial event in the evolution of the unusual polyandrous relationships occurring in the Tasmanian Native Hen (*Tribonyx mortierii*). The prevailing view, however, seems to be that the mating relationships occurring in a population result from the distribution and availability of resources and the individuals' relative abilities in controlling these resources (Orians 1961, 1969; Verner 1964; Verner and Willson 1966; Lack 1968; Selander 1972; Jenni 1974; Pitelka et al. 1974; Wittenberger 1976, 1979; Altmann et al. 1977; Emlen and Oring 1977; Graul et al. 1977; Halliday 1978; Borgia 1979; Wittenberger and Tilson 1980).

In this paper I discuss the influence of a population's age structure and sex-ratio on the evolution of monogamy.

DEFINITIONS

The terms monogamy, polygyny, polyandry, and polygyny-polyandry (e.g., the Greater Rhea *Rhea americana,* Bruning 1973) should not usually be applied to species or populations. Such use can be misleading. For example, Verner and Willson (1969) defined polygyny as the mating system in which at least 5% of the males had two or more mates simultaneously. Although this definition is acceptable for the purposes of Verner and Willson, it is not acceptable for thinking about the evolution of mating behavior because it obscures the fact that within a "polygynous species" as many as 95% of the males and 100% of the females may have no more than a single mate.

The terms are better used to describe the relationships between males and females in distributing their gametes. No reference is made to the length or quality of the pair-bond (see also Gowaty 1981a). For my purposes, the terms refer to the number of mates an individual has during a single reproductive cycle. In a monogamous relationship each sex has a single mate. In polygyny, males have more than one mate. In polyandry, females have more than one mate. And, in polygyny-polyandry, each sex has more than one mate. According to my definitions, the number of mates each individual has is the only criterion for recognizing the relationship. Including other criteria in the definitions leads to a more complex classification and may compound two or more separate evolutionary problems. For example, if polygyny and promiscuity are distinguished by the relative length of the pair-bonds between the mated individuals, then we may tend to think in terms of two different mating systems. If instead we define polygyny as I have done, then we can think of "polygyny with long-term pair-bonds" and "polygyny with brief pair-bonds" and realize that we may be dealing with two different problems, the evolution of polygyny and the evolution of pair-bonds.

A Demographic Equation

If we can assume that populations have structure, that is, they consist of males and females of particular ages (whether or not we can count them), and if we can assume average values of age-specific mortality and fecundity, then we can use Lotka's equation,

$$1 = \Sigma \lambda_x \mu_x e^{-\rho x} \tag{1}$$

[where λ_x is the probability of an individual's surviving from birth to age class x, μ_x is the average number of progeny born to or sired by individuals of age class x, ρ is the rate of increase of the population, e is the base of the natural logarithms, and x is the age class (Murray and Gårding 1984)], to estimate values of ρ for subpopulations of individuals with particular traits. We can then compare these values in evaluating the relative success of individuals with alternative traits. For example, we could compare ρ values for birds beginning to breed at age 1 and others beginning to breed at age 2. If the former trait has the greater ρ, then the latter trait should eventually decline to near extinction.

In thinking about the evolution of mating systems, it is useful to calculate the average annual replacement fecundity (μ_r) of individuals making up subpopulations with different life history characteristics, in particular, different survivorship schedules and ages of first breeding. Given several simplifying assumptions (Murray 1979), the average annual replacement fecundity is given by

$$\mu_r = 2 \Big/ \sum_{\alpha}^{\omega} \lambda_x, \tag{2}$$

where α is the age class of first reproduction and ω is the age class of last reproduction. For a given survivorship schedule, the average annual replacement fecundity increases as the age of first breeding is postponed (Table 1). Individuals

TABLE 1

SURVIVORSHIP, AGES OF FIRST BREEDING, AND AVERAGE ANNUAL FECUNDITY OF A HYPOTHETICAL POPULATION

Age	Age class (x)	Probability of surviving to age class x (λ_x)	Age class at first reproduction (α)	Average annual replacement fecundity (μ_r)	Average annual fecundity (μ_a)
0	1	1.000	1	1.455	2
1	2	0.200	2	5.333	6
2	3	0.100	3	11.429	12
3	4	0.050	4	26.667	27
4	5	0.025	5	80.000	80
5	6	0.000	—	—	—

with below average annual replacement fecundities are being selected against, and those with above average annual replacement fecundities are favored.

POPULATION SIMULATION

Since the publication of Williams's (1966) book, *Adaptation and Natural Selection,* evolutionists have placed much emphasis on the individual. Many evolutionary "hypotheses" have the form, "if individuals benefit from behaving in a particular way, then that behavior evolves." These are often no more than stories, and, as pointed out by Gould and Lewontin (1979), the only test of many of the stories told by evolutionists is their consistency with the theory of natural selection. But, evolution is a population phenomenon. It is the change in frequencies of alternative traits within a population over the course of time. Trait frequencies change by natural selection when individuals with one trait are more successful in surviving and reproducing than individuals with alternative traits. The result is a population of individuals with the favorable trait. Thus, in considering the evolution of a trait that seems beneficial to individuals, we must determine whether a population of individuals with that trait could exist. This is especially so with regard to behavioral traits involving interaction with other members of the population, such as mating behavior. In fact, as we shall see, a population's age structure and sex-ratio place constraints on the kind(s) of mating relationship(s) that can occur within a population.

We may see how a population's structure can affect the evolution of monogamy by considering a hypothetical population of 20 males and 20 females, all sexually mature and ready to breed, as representative of any population with a sex-ratio of sexually mature males and females of 1.0. If the males are intrasexually aggressive and establish territories, they may divide the available habitat in such a way that only 16 are established as territory holders and 4 are not (Fig. 1). What kind(s) of mating relationship(s) can occur?

Clearly, only 16 females can be mated to monogamous males. What happens to the remaining 4 females? They seem to have four options: (1) not to mate at all, ever, (2) mate with non-territorial males in monogamous relationships, (3) mate with already mated males in polygynous relationships, or (4) wait until the following year or later and breed in monogamous relationships. If these are alternative traits that can be acted upon by natural selection, that is, these traits differ genetically, then which one prevails?

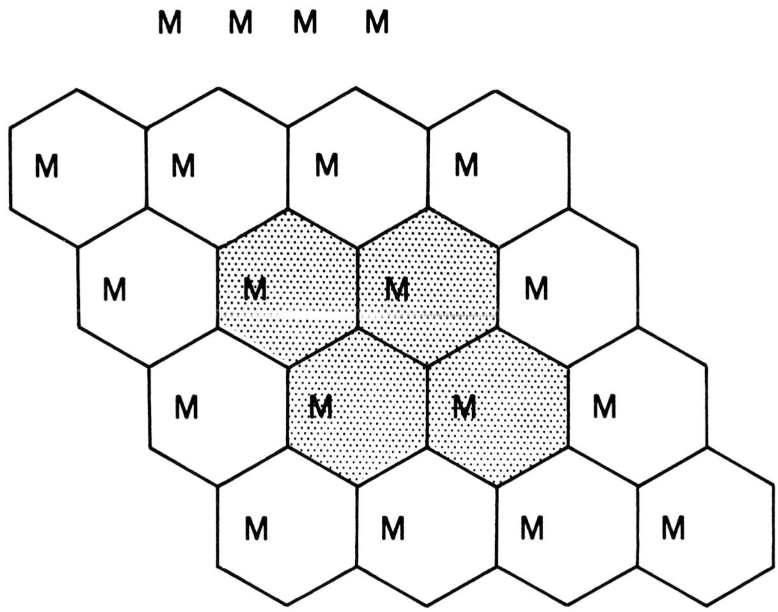

FIG. 1. Territories in a hypothetical population. If 16 of 20 sexually mature males are able to establish territories, a few of which are better quality (shaded) than the others (unshaded), then only 16 females can be mated to a monogamous male with a territory. The mating relationships in this population depend upon the sex-ratio and upon whether the more numerous sex can afford to wait until a later year for a monogamous relationship rather than upon differences in the quality of territories.

Each of these possibilities has costs. Any genetic combination responsible for traits resulting in non-breeding behavior at all ages (option 1) must surely be severely selected against. Females rarely mate with non-territorial males (option 2), presumably because such behavior has a high cost. Mating with already mated males (option 3) has a cost in reduced assistance in parental care from their mates, whereas waiting to breed (option 4) has a cost in greater annual reproductive effort compared with that for breeding earlier (Table 1). Nevertheless, females often do mate with already mated males in polygynous relationships or wait and breed at a later time in monogamous relationships.

In summary, options (1) and (2) are rare and seemingly maladaptive. Options (3) and (4) occur frequently, and each is presumably adaptive under particular conditions. These last two options require further evaluation.

Although recent theory regards the distribution and abundance of resources and the relative abilities of individuals to control resources as the important factors in determining mating relationships (see references cited above), I believe (Murray 1984) that a population's age structure and sex-ratio affect the evolution of territorial behavior and mating relationships (see Murray 1979, for a demographic theory on the evolution of clutch-size).

NUMERICAL EXAMPLES

We can evaluate the constraining influence of a population's age structure and sex-ratio on the evolution of mating relationships by examining the options open

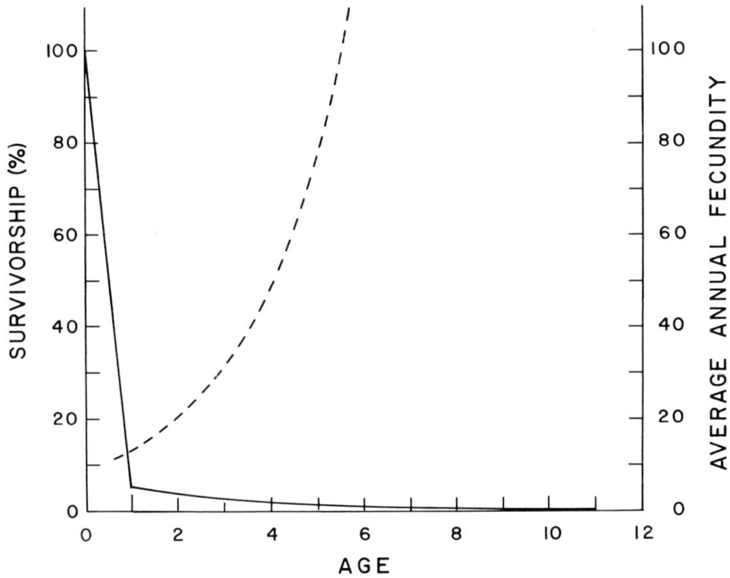

FIG. 2. Survivorship and average annual fecundity in the Prairie Warbler. Survivorship (solid line) is drawn from data in Nolan (1978). The average annual replacement fecundity (dashed line) of females beginning to breed at a given age is calculated from equation (2).

to female Prairie Warblers (*Dendroica discolor*) and female Laysan Albatrosses (*Diomedea immutabilis*) by applying Lotka's equation to survivorship data. The data discussed below for the Prairie Warbler are from Nolan (1978), and those for the Laysan Albatross are from Fisher (1975) and Van Ryzin and Fisher (1976).

In the Prairie Warbler, annual mortality is high, with less than 6% surviving the first year (Fig. 2). With this survivorship schedule, females that begin breeding at one year of age must produce, on the average, 14 eggs in every year of their reproductive life if they are to replace themselves and their mates (Fig. 2). Females that begin breeding at two years of age, however, must produce, on average, 21 eggs in every year of their reproductive life if they are to replace themselves and their mates (Fig. 2), much less do better than the females beginning to breed earlier (assuming for simplicity that survivorship is unaffected by the differences in fecundity [see Discussion]).

I suggest that 21 eggs per year could in fact impose much too high a cost in reproductive effort on female Prairie Warblers, requiring females to begin breeding in their first year. As the sex-ratio in the Prairie Warbler favors males at the beginning of the breeding season, the Prairie Warbler population has some polygynous matings, about 5% at the beginning of the breeding season. My conclusion is that all the benefits for females of mating with monogamous males are irrelevant if females must breed in their first year and if there are not enough males to provide each female with a monogamous mate.

By contrast, survivorship of the Laysan Albatross is high (Fig. 3). With this survivorship schedule, females that lay a single egg in every year of reproductive life need not lay their first egg until the age of 15 (Fig. 3) if they are to replace themselves. However, Laysan Albatrosses lay less than one egg each year, and

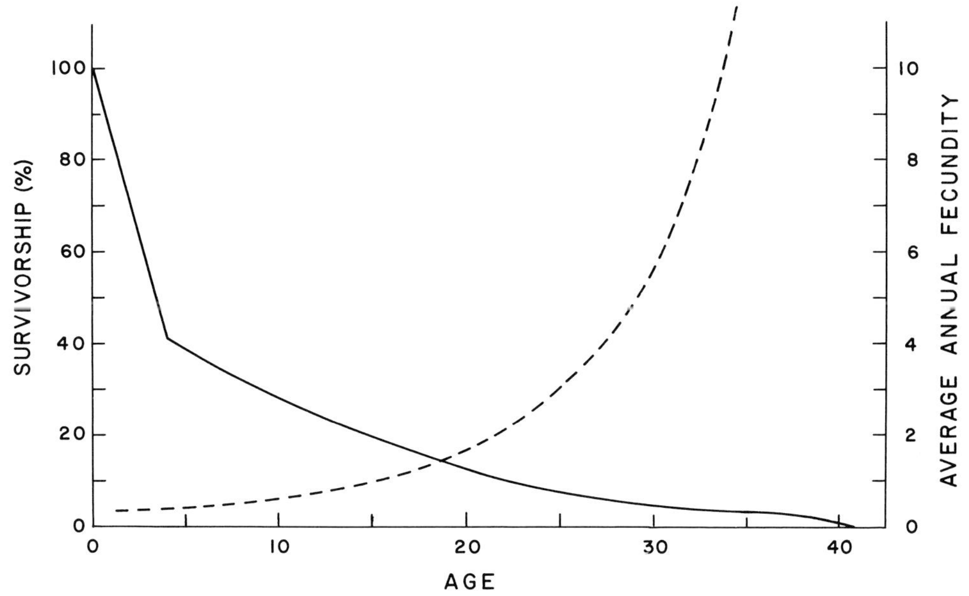

FIG. 3. Survivorship and average annual fecundity in the Laysan Albatross. Survivorship (solid line) is drawn from data in Fisher (1975). The average annual replacement fecundity (dashed line) of females beginning to breed at a given age is calculated from equation (2).

thus they must begin breeding at an earlier age. For example, females that lay, on the average, 2 eggs every 3 years must begin breeding by the age of 11. Female Laysan Albatrosses, then, are in no hurry to begin breeding. They can wait for a monogamous relationship, and this they do. The average age of first breeding of female Laysan Albatrosses is 8.9 years, whereas the average age of first breeding in males is 8.4 years. I conclude that the demographics of the Laysan Albatross allow the evolution of obligate monogamy, whereas the demographics of the Prairie Warbler do not.

DISCUSSION

The examples of the Prairie Warbler and Laysan Albatross provide only simple illustrations of how a population's age structure and sex-ratio can influence the mating relationships between its members. In fact, calculating specific values for various (sub)population parameters is much more complex. Each combination of traits is likely to have a different survivorship schedule. For example, postponing breeding in the first year reduces mortality through the second year, compared with that for birds breeding in their first year, and the greater reproductive effort once breeding begins must increase mortality relative to that for birds beginning to breed at an earlier age with a smaller annual reproductive effort. Thus, comparing quantitatively the ρ values of alternative traits requires knowing the survivorship schedules of the groups exhibiting alternative traits. What few survivorship schedules exist, however, are composites, including all the individuals (at least of one sex) in the sample population.

Even though we are never likely to have the quantitative demographic data to evaluate the success of alternative traits, we can at least think about how different patterns of mortality affect the adult sex-ratio and the probabilities that males and females have in obtaining mates. It happens that the potential number of combinations of male survivorship, female survivorship, and ages of first and last reproduction is almost infinite, but it also happens that there are only a few general patterns. These have been presented and explored in detail elsewhere (Murray 1984).

It is tempting to believe that the cause-and-effect relationship is just the reverse, viz., that differential survivorship of the sexes and the sex-ratio are consequences of selection favoring particular mating systems. It is certainly conceivable that one or the other sex may suffer greater mortality because of differential parental effort. For example, if one sex does most of the incubation, and if incubators suffer greater mortality than non-incubators, then differences in parental behavior can effect the sex-ratio. No doubt, one factor in determining differential survivorship of the sexes and, therefore, the sex-ratio is the different roles the sexes may have in reproduction. But these roles vary greatly, and therefore one cannot predict the direction, much less the magnitude, of the sex-ratio. For example, among monogamous species, females of the Red-billed Gull (*Larus novaehollandiae*; Mills 1973) and Black-legged Kittiwake (*Rissa tridactyla*; Coulson and Wooller 1976) survive better than the males, whereas in the Yellow-eyed Penguin (*Megadyptes antipodes*; Richdale 1951) and the Adélie Penguin (*Pygoscelis adeliae*; Ainley and DeMaster 1980) males survive better than females. In these species, at least, monogamy seems not to have produced a consistent pattern of sex-ratios.

It remains to be demonstrated that polygyny necessarily leads to greater mortality among males and polyandry to greater mortality among females, thus accounting for the sex-ratios often associated with these mating systems. Theoretically, at least, any mating system can occur with any sex-ratio (Murray 1984), suggesting that mating systems cannot predict the sex-ratio, even qualitatively. What matters is how the sex-ratio and the female's annual fecundity influence the evolution of intrasexual aggression and the ages of first reproduction of males and females (Murray 1984).

Analysis of the data on the Prairie Warbler and the Laysan Albatross indicates that when females have a choice between mating with an already mated male now and mating with an unmated male later, obligate monogamy (i.e., polygamy very rare) evolves when the sexes can afford to wait for a monogamous mate. When females cannot afford to wait, they join already mated males in polygynous relationships. Males with only one mate in such populations are facultatively monogamous. Although not presented in this paper, a similar analysis can be made for the options open to males, and when males cannot afford to wait for monogamous mates because of a shortage of females, they join other males in polyandrous relationships. Females with only one mate in such populations are facultatively monogamous.

Acknowledgments

I thank Patricia A. Gowaty, Douglas W. Mock, Lewis W. Oring, William M. Shields, and an anonymous reviewer for their comments on the manuscript.

SUMMARY

After defining the various mating systems (monogamy, polygyny, polyandry, and polygyny-polyandry) as relationships between mated individuals, and after suggesting that the probabilities of successful reproduction by individuals can be calculated with Lotka's equation, a hypothetical population with an adult sex-ratio of unity is considered to show how a population's age structure and sex-ratio affect the kind(s) of mating relationship(s) that can occur in a population. Often females have a choice between mating with an already mated male in a polygynous relationship now or waiting until a later breeding season and mating with an unmated male in a monogamous relationship. Although polygynous relationships incur a cost to females of reduced parental care from their mates, waiting incurs a cost in increased annual fecundity. These alternatives are evaluated for the Prairie Warbler and Laysan Albatross. This evaluation indicates that obligate monogamy can evolve only when females can afford to wait. Otherwise, females accept polygynous relationships.

LITERATURE CITED

AFTON, A. D. 1979. Time budget of breeding Northern Shovelers. Wilson Bull. 91:42–49.

AFTON, A. D. 1983. Male and female strategies for reproduction in Lesser Scaup. Unpub. Ph.D. dissertation, University of North Dakota, Grand Forks, North Dakota.

AFTON, A. D. 1984. Influence of age and time on reproductive performance of female Lesser Scaup. Auk 101:255–265.

AINLEY, D. G., AND D. P. DEMASTER. 1980. Survival and mortality in a population of Adélie Penguins. Ecology 61:522–530.

ALATALO, R. V., L. GUSTAFSSON, AND A. LUNDBERG. 1984. High frequency of cuckoldry in the Pied and Collared flycatchers. Oikos 42:41–47.

ALEXANDER, R. D. 1974. The evolution of social behavior. Ann. Rev. Ecol. Syst. 5:324–383.

ALEXANDER, R. D., AND G. BORGIA. 1979. On the origin and basis of the male-female phenomenon. Pp. 412–440, *In* M. S. Blum and N. A. Blum (eds.), Sexual Selection and Reproductive Competition in Insects. Academic Press, New York.

ALTMANN, S. A., S. S. WAGNER, AND S. LENINGTON. 1977. Two models for the evolution of polygyny. Behav. Ecol. Sociobiol. 2:397–410.

ANDERSON, M. G. 1984. Parental investment and pairbond behavior among Canvasback Ducks (*Aythya valisineria, Anatidae*). Behav. Ecol. Sociobiol. 15:81–90.

ANDERSON, M. G. 1985. Social behavior of the Canvasback (*Aythya valisineria*): male and female strategies of reproduction. Unpub. Ph.D. dissertation, University of Minnesota, Minneapolis, Minnesota.

ANDERSON, M. G., R. D. SAYLER, AND A. D. AFTON. 1980. A decoy trap for diving ducks. J. Wildl. Manage. 44:217–219.

ANDERSSON, M. 1983. Brood parasitism within species. *In* C. J. Barnard (ed.), Producers and Scroungers: Strategies of Exploitation and Parasitism. Croom Helm, London, England.

ANDERSSON, M., AND M. G. ERIKSSON. 1982. Nest parasitism in Goldeneyes, *Bucephala clangula*: some evolutionary aspects. Am. Nat. 120:1–16.

ARAK, A. 1983. Male-male competition and mate choice in anuran amphibians. Pp. 181–210, *In* P. Bateson (ed.), Mate Choice. Cambridge University Press, Cambridge, England.

ARMSTRONG, E. A. 1947. Bird Display and Behaviour. Lindsay Drummond, London, England.

ARNOLD, S. J. 1983. Sexual selection: the interface of theory and empiricism. Pp. 67–108, *In* P. Bateson (ed.), Mate Choice. Cambridge University Press, Cambridge, England.

ASHCROFT, R. E. 1976. A function of the pairbond in the Common Eider. Wildfowl 27:101–105.

BAILEY, R. O., N. R. SEYMOUR, AND G. R. STEWART. 1978. Rape behavior in Blue-winged Teal. Auk 95:188–190.

BALL, I. J., P. G. H. FROST, W. R. SIEGFRIED, AND F. MCKINNEY. 1978. Territories and local movements of African Black Ducks. Wildfowl 29:61–79.

BANCROFT, J. 1978. The relationship between hormones and sexual behavior in humans. Pp. 111–143, *In* J. Hutchinson (ed.), Biological Determinants of Sexual Behavior. Wiley, London, England.

BARASH, D. P. 1976. Male response to apparent female adultery in the Mountain Bluebird (*Sialia currucoides*): an evolutionary interpretation. Am. Nat. 110:1097–1101.

BARASH, D. P. 1977. Sociobiology of rape in Mallards (*Anas platyrhynchos*): responses of the mated male. Science 197:788–789.

BARRETT, J. 1973. Breeding behavior of captive Mallards. Unpub. M.S. thesis, University of Minnesota, Minneapolis, Minnesota.

BARTONEK, J. C., AND C. W. DANE. 1964. Numbered nasal discs for waterfowl. J. Wildl. Manage. 28:688–692.

BATEMAN, A. J. 1948. Intrasexual selection in *Drosophila*. Heredity 2:349–368.

BATES, H., AND R. BUSENBARK. 1970. Finches and Soft-billed Birds. T. F. H. Publications, Neptune City, New Jersey.

BATESON, P. P. G. 1978. Sexual imprinting and optimal outbreeding. Nature 273:659–660.

BATESON, P. P. G. 1980. Optimal outbreeding and the development of sexual preferences in Japanese Quail. Z. Tierpsychol. 53:231–244.

BATESON, P. P. G. 1982. Behavioural development and evolutionary processes. Pp. 133–151, *In* King's College Sociobiology Group (ed.), Current Problems in Sociobiology. Cambridge University Press, Cambridge, England.

BATESON, P. P. G., W. LOTWICK, AND D. K. SCOTT. 1980. Similarities between the faces of parents and offspring in Bewick's Swan and the differences between mates. J. Zool. 191:61–74.

BEECHER, M. D., AND I. M. BEECHER. 1979. Sociobiology of Bank Swallows: reproductive strategy of the male. Science 205:1282–1285.

BEEHLER, B. 1983a. The behavioral ecology of four birds of paradise. Unpub. Ph.D. dissertation, Princeton University, Princeton, New Jersey.

BEEHLER, B. 1983b. Frugivory and polygamy in birds of paradise. Auk 100:1–12.

BEEHLER, B. 1983c. Notes on the behaviour and ecology of Macgregor's Bird of Paradise. Emu 83: 28–30.

BEEHLER, B., AND S. G. PRUETT-JONES. 1983. Display dispersion and diet of male birds of paradise: a comparison of eight species. Behav. Ecol. Sociobiol. 13:229–238.

BELLROSE, F. C. 1976. Ducks, Geese and Swans of North America. Stackpole, Harrisburg, Pennsylvania.

BELLROSE, F. C., T. G. SCOTT, A. S. HAWKINS, AND J. B. LOW. 1961. Sex ratios and age ratios in North American ducks. Ill. Nat. Hist. Surv. Bull. 27:391–474.

BERTRAM, B. C. R. 1979. Ostriches recognize their own eggs and discard others. Nature 279:233–234.

BEZZEL, E. 1964. Zum vorkommen von polygynie bei Enten. Die Vogelwelt 85:39–43.

BEZZEL, E. 1969. Die Tafelente. Neue Brehm Buecherei. A Ziemsen Verlag, Lutherstadt, Germany.

BLUHM, C. K. 1981. Environmental controls of breeding in captive Canvasback ducks (*Aythya valisineria*). Unpub. Ph.D. dissertation, University of Minnesota, Minneapolis, Minnesota.

BLUM, M. S., AND N. A. BLUM (EDS.). 1979. Sexual selection and reproductive competition in insects. Academic Press, New York.

BOAG, P. T., AND P. R. GRANT. 1981. Intense natural selection in a population of Darwin's finches (Geospizinae) in the Galapagos. Science 214:82–85.

BOOKHOUT, T. A. 1979. Waterfowl and Wetlands—an Integrated Review. La Crosse Printing Co., La Crosse, Wisconsin.

BORGIA, G. 1979. Sexual selection and the evolution of mating systems. Pp. 19–80, *In* M. S. Blum and N. A. Blum (eds.), Sexual Selection and Reproductive Competition in Insects. Academic Press, New York.

BOSSEMA, I., AND J. P. KRUIJT. 1982. Male activity and female mate acceptance in the Mallard (*Anas platyrhynchos*). Behaviour 79:313–324.

BRADBURY, J. W. 1981. The evolution of leks. Pp. 138–169, *In* R. D. Alexander and D. Tinkle (eds.), Natural Selection and Social Behavior. Chiron Press, New York.

BRAITHWAITE, L. W. 1976a. Breeding seasons of waterfowl in Australia. Proc. 16th Int. Ornithol. Congr. Pp. 235–247.

BRAITHWAITE, L. W. 1976b. Environment and timing of reproduction and flightlessness in two species of Australian ducks. Proc. 16th Int. Ornithol. Congr. Pp. 486–501.

BRAITHWAITE, L. W., AND H. J. FRITH. 1969. Waterfowl in an inland swamp in New South Wales. III. Breeding. CSIRO Wildlife Research 14:65–109.

BRAITHWAITE, L. W., AND F. I. NORMAN. 1974. The 1972 open season on waterfowl in South-eastern Australia. CSIRO Aust. Div. Wildl. Res. Tech. Paper No. 29:1–37.

BRAITHWAITE, L. W., AND F. I. NORMAN. 1976. The 1973 and 1974 open seasons on waterfowl in South-eastern Australia. CSIRO Div. Wildl. Res. Tech. Memo No. 11:1–58.

BRAITHWAITE, L. W., AND F. I. NORMAN. 1977. The 1975 and 1976 open seasons on waterfowl in South-eastern Australia. CSIRO Aust. Div. Wildl. Res. Tech. Memo No. 13:1–62.

BRAY, O. E., J. J. KENNELLY, AND J. L. GUARINO. 1975. Fertility of eggs produced on territories of vasectomized Red-winged Blackbirds. Wilson Bull. 87:187–195.

BROOKE, M. DE L. 1978. Some factors affecting the laying date, incubation and breeding success of the Manx Shearwater, *Puffinus puffinus*. J. Anim. Ecol. 47:477–495.

BROWN, J. L. 1969. Territorial behavior and population regulation in birds: a review and re-evaluation. Wilson Bull. 81:293–329.

BROWN, C. F. 1984. Laying eggs in a neighbor's nest: benefit and cost of colonial nesting in swallows. Science 224:518–519.

BRUNING, D. F. 1973. The Greater Rhea chick and egg delivery route. Natural History 82:68–75.

BRUNING, D. F. 1974. Social structure and reproductive behavior of the Greater Rhea. Living Bird 13:251–294.

BUECHNER, H. K., AND R. SCHLOETH. 1965. Ceremonial mating behavior in Uganda Kob (*Adenota kob thomase* Neuman). Z. Tierpsychol. 22:209–225.

BULLOUGH, W. S. 1942. The reproductive cycle of the British and continental races of the Starling *Sturnus vulgaris*. Trans. Royal Soc. Lond. B231:165–246.

BURGHARDT, G. M. 1970. Defining "communication." Pp. 5–18, *In* G. W. Johnston, D. G. Houlton and A. Turk (eds.), Advances in Chemoreception, Vol. 1. Appleton-Century-Crofts, New York.

BURLEY, N. 1977a. Parental investment, mate choice, and mate quality. Proc. Nat. Acad. Sci. U.S.A. 74:3476–3479.

BURLEY, N. 1977b. Mate choice and sexual selection in the Pigeon, *Columba livia*. Ph.D. dissertation, University of Texas, Austin, Texas.

BURLEY, N. 1981a. The evolution of sexual indistinguishability. Pp. 121–137, *In* R. D. Alexander and D. W. Tinkle (eds.), Natural Selection and Social Behavior. Chiron Press, New York.

BURLEY, N. 1981b. Mate choice by multiple criteria in a monogamous species. Am. Nat. 117:515–528.

BURLEY, N., G. KRANTZBERG, AND P. RADMAN. 1982. Influence of colour-banding on the conspecific preferences of Zebra Finches. Anim. Behav. 30:444–445.

BURLEY, N., AND N. MORAN. 1979. The significance of age and reproductive experience in the mate preferences of feral Pigeons, *Columba livia*. Anim. Behav. 27:686–698.

BURNS, J. T., K. CHENG, AND F. MCKINNEY. 1980. Forced copulation in captive Mallards. I. Fertilization of eggs. Auk 97:875–879.

BUTTERFIELD, P. A. 1970. The pair bond in the Zebra Finch. Pp. 249–278, *In* J. H. Crook (ed.), Social Behaviour in Birds and Mammals. Academic Press, London, England.

CADE, W. H. 1984. Genetic variation underlying sexual behavior and reproduction. Amer. Zool. 24: 355–366.

CAMPBELL, B. (ED.). 1972. Sexual Selection and the Descent of Man, 1871–1971. Aldine, Chicago, Illinois.

CAREY, M., AND V. NOLAN, JR. 1979. Population dynamics of Indigo Buntings and the evolution of avian polygyny. Evolution 33:1180–1192.

CARSON, H. L. 1968. The population flush and its genetic consequences. Pp. 123–137, *In* R. C. Lewontin (ed.), Population Biology and Evolution. Syracuse University Press, Syracuse, New York.

CHENG, K. M., J. T. BURNS, AND F. MCKINNEY. 1982. Forced copulation in captive Mallards (*Anas platyrhynchos*). II. Temporal factors. Anim. Behav. 30:695–699.

CHENG, K. M., J. T. BURNS, AND F. MCKINNEY. 1983. Forced copulation in captive Mallards: III. Sperm competition. Auk 100:302–310.

CHENG, K. M., R. N. SHOFFNER, R. E. PHILLIPS, AND F. B. LEE. 1978. Mate preference in wild and domesticated (game-farm) Mallards: I. Initial preference. Anim. Behav. 26:996–1003.

CHENG, K. M., R. N. SHOFFNER, R. E. PHILLIPS, AND F. B. LEE. 1979. Mate preference in wild and domesticated (game-farm) Mallards: II. Pairing success. Anim. Behav. 27:417–425.

CLARK, A. 1969. The breeding of the Hottentot Teal. Ostrich 40:33–36.

CLUTTON-BROCK, T. H., F. E. GUINESS, AND S. D. ALBON. 1982. Red Deer: Behavior and Ecology of Two Sexes. University Chicago Press, Chicago, Illinois.

CONOVER, W. J. 1971. Practical Nonparametric Statistics. John Wiley and Sons, New York.

COOKE, F., M. A. BOUSFIELD, AND A. SADURA. 1982. Mate change and reproductive success in the Lesser Snow Goose. Condor 83:322–327.

COOKE, F., G. H. FINNEY, AND R. F. ROCKWELL. 1976. Assortative mating in Lesser Snow Geese. Behav. Genetics 6:127–140.

COOPER, W. S., AND J. FORSHAW. 1977. The Birds of Paradise and Bowerbirds. Collins Press, Sydney, Australia.

COULSON, J. C. 1966. The influence of the pair-bond on the breeding biology of the Kittiwake Gull, *Rissa tridactyla*. J. Anim. Ecol. 35:269–279.

COULSON, J. C. 1972. The significance of the pair-bond in the Kittiwake. Proc. 15th Int. Ornithol. Congr. Pp. 424–433.

COULSON, J. C., AND R. D. WOOLLER. 1976. Differential survival rates among breeding Kittiwake gulls *Rissa tridactyla* (L.). J. Anim. Ecol. 45:205–213.

COWAN, I. MCT., AND J. HATTER. 1952. A trap technique for the capture of diving waterfowl. J. Wildl. Manage. 16:438–441.

COX, C. R., AND B. J. LEBOEUF. 1977. Female incitation of male competition: a mechanism in sexual selection. Am. Nat. 111:317–335.

CRAMP, S., AND K. E. L. SIMMONS (EDS.). 1977. Handbook of the birds of Europe, the Middle East, and North Africa. The Birds of the Western Palearctic Vol. I. Oxford University Press, Oxford, England.

CROME, F. H. J. 1975. The ecology of fruit pigeons in tropical northern Queensland. Austr. Wild. Res. 2:155–185.

CROOK, J. H. 1964. The evolution of social organization and visual communication in the weaver birds (Ploceinae). Behaviour Suppl. 10:1–178.

CROOK, J. H. 1965. The adaptive significance of avian social organisation. Symp. Zool. Soc. London 14:181–218.

CUMMINGS, G. E., AND O. H. HEWITT. 1964. Capturing waterfowl and marsh birds at night with light and sound. J. Wildl. Manage. 28:120–126.

DALY, M., AND M. WILSON. 1983. Sex, Evolution, and Behavior. Willard Grant Press, Boston, Massachusetts.

DARWIN, C. 1871. The Descent of Man, and Selection in Relation to Sex. John Murray, London, England.

DAVIS, J. W. F. 1976. Breeding success and experience in the Arctic Skua, *Stercorarius parasiticus* (L.). J. Anim. Ecol. 45:531–536.

DAWKINS, R. 1980. Good strategy or evolutionarily stable strategy? Pp. 331–367, *In* G. W. Barlow and J. Silverberg (eds.), Sociobiology: Beyond Nature/Nurture? AAAS Symp. 35. Westview Press, Boulder, Colorado.

DEAN, W. R. J. 1978. Moult seasons of some Anatidae in the western Transvaal. Ostrich 49:76–84.

DEAN, W. R. J., AND D. M. SKEAD. 1977. The sex ratio in Yellow-billed Duck, Red-billed Teal and Southern Pochard. Ostrich Suppl. 12:82–85.

DELACOUR, J. 1954–1964 (4 Volumes). The Waterfowl of the World. Country Life Ltd., London, England.

DELACOUR, J., AND E. MAYR. 1945. The family Anatidae. Wilson Bull. 57:2–55.

DEMENT'EV, G. P., AND N. A. GLADKOV (EDS.). 1952. Birds of the Soviet Union Vol. 4. Israel Program for Scientific Translation, Jerusalem, Israel.

DERRICKSON, S. R. 1977. Aspects of breeding behavior in the Pintail (*Anas acuta*). Unpub. Ph.D. dissertation, University of Minnesota, Minneapolis, Minnesota.

DERRICKSON, S. R. 1978. Mobility of breeding pintails. Auk 95:104–114.

DHARMAKUMARSINHJI, K. S. 1943. Notes on the breeding of the Empress of Germany Bird of Paradise. Zoologica 28:139–144.

DONAGHEY, R. H. 1981. The ecology and evolution of bowerbird mating systems. Unpub. Ph.D. dissertation, Monash University, Clayton, Victoria, Australia.

DOTY, H. A., AND R. J. GREENWOOD. 1974. Improved nasal saddles for mallards. J. Wildl. Manage. 38:938–939.

DOTY, H. A., D. L. TRAUGER, AND J. R. SERIE. 1984. Renesting by female Canvasbacks in southwestern Manitoba. J. Wildl. Manage. 48:581–584.

DUNBAR, R. I. M. 1982. Intraspecific variations in mating strategy. Pp. 385–431, *In* P. P. G. Bateson and P. Klopfer (eds.), Perspectives in Ethology. Vol. 5. Plenum, New York.

DWYER, T. J. 1974. Social behavior of breeding Gadwall in North Dakota. Auk 91:375–386.

DWYER, T. J. 1975. Time budget of breeding Gadwalls. Wilson Bull. 87:335–343.

DZUBIN, A. 1969. Assessing breeding populations of ducks by ground counts. Saskatoon Wetlands Seminar, Can. Wildl. Serv. Rept. 6:178–230.

EBERHARDT, L. E., AND A. B. SARGEANT. 1977. Mink predation on prairie marshes during the waterfowl breeding season. Pp. 32–43, *In* R. L. Phillips and C. Jonkel (eds.), Proc. 1975 Predation Symposium. Montana Forest and Conserv. Exper. Sta., Missoula, Montana.

EMLEN, S. T., AND L. W. ORING. 1977. Ecology, sexual selection and the evolution of mating systems. Science 197:215–223.

ERICKSON, C. J. 1978. Sexual affiliation in animals: pair bonds and reproductive strategies. Pp. 696–725, *In* J. S. Hutchinson (ed.), Biological Determinants of Sexual Behavior. Wiley, New York.

ERICKSON, C. J., AND P. G. ZENONE. 1976. Courtship differences in male Ring Dove, avoidance of cuckoldry? Science 197:1352–1354.

ERICKSON, R. C. 1948. Life history and ecology of the Canvas-back (*Nyroca valisineria*) (Wilson), in southeastern Oregon. Unpub. Ph.D. dissertation, Iowa State College, Ames, Iowa.

EVANS, P. G. H. 1982. Population genetics of the European Starling *Sturnus vulgaris*. Unpub. Ph.D. dissertation. University of Oxford, Oxford, England.

EVERITT, C. 1965. Breeding of the Magnificent Bird of Paradise. Avicul. Mag. 71:146–148.

FALCONER, D. S. 1981. Introduction to Quantitative Genetics. 2nd ed., Longman, New York.

FIALA, K. L. 1981. Reproductive cost and the sex ratio in Red-winged Blackbirds. Pp. 198–214, *In* R. D. Alexander and D. W. Tinkle (eds.), Natural Selection and Social Behavior. Chiron Press, New York.

FISHER, H. I. 1975. The relationship between deferred breeding and mortality in the Laysan Albatross. Auk 92:433–441.

FISHER, R. A. 1930. The Genetical Theory of Natural Selection. Clarendon Press, Oxford, England.

FORD, N. L. 1983. Variations in mate fidelity in monogamous birds. Pp. 329–356, *In* R. F. Johnston (ed.), Current Ornithology, Vol. 1. Plenum, New York.

FOSTER, M. S. 1978. Total frugivory in tropical birds: a reappraisal. Trop. Ecol. 19:131–154.

FOSTER, M. S., AND R. W. MCDIARMID. 1983. Nutritional value of the aril of *Trichilia cuneata:* a bird-dispersed fruit. Biotropica 15:26–31.

FRAZZETTA, T. 1975. Complex Adaptations in Evolving Populations. Sinauer Assoc., Sunderland, Massachusetts.

FRITH, H. J. 1959. The ecology of wild ducks in inland New South Wales. IV. Breeding. CSIRO Wildlife Research 4:156–181.

FRITH, H. J. 1967. Waterfowl in Australia. East-West Center Press, Honolulu, Hawaii.

FROST, P. G. H. 1980. Fruit-frugivore interactions in a South African coastal dune forest. Proc. 17th Int. Ornithol. Congr., pp. 1179–1184.

FUJIOKA, M., AND S. YAMAGISHI. 1981. Extramarital and pair copulations in the Cattle Egret. Auk 98:134–144.

GARSON, P. J., J. L. DUNN, C. J. WALTON, AND P. A. SHAW. 1980. Stimuli eliciting courtship from domesticated Zebra Finches. Anim. Behav. 28:1184–1187.

GHISELIN, M. T. 1974. The Economy of Nature and the Evolution of Sex. University of California Press, Berkeley, California.

GIESEL, J. T., AND E. E. ZETTLER. 1980. Genetic correlations of life history parameters and certain fitness indices in *Drosophila melanogaster*: r(m), r(s), diet breadth. Oecologia 47:299–302.

GILLIARD, E. T. 1969. The Birds of Paradise and Bower Birds. Natural History Press, Garden City, New York.

GLADSTONE, D. E. 1979. Promiscuity in monogamous colonial birds. Am. Nat. 144:545–557.

GOODWIN, D. 1982. Estrildid Finches of the World. Cornell University Press, Ithaca, New York.

GORSKI, R. A. 1979. The neuroendocrinology of reproduction: an overview. Biol. Reprod. 20:111–127.

GOSS, R. J. 1965. The functional demand theory of growth regulation. Pp. 444–451, *In* V. Kiortsis and H. A. L. Trampusch (eds.), Regeneration in Animals and Related Problems. North-Holland, Amsterdam, Netherlands.

GOULD, J. L. 1982. Ethology: the Mechanisms and Evolution of Behavior. W. W. Norton, New York.

GOULD, S. J., AND R. C. LEWONTIN. 1979. The spandrels of San Marco and the Panglossian paradigm: a critique of the adaptationist programme. Proc. R. Soc. London B 205:581–598.

GOWATY, P. A. 1980. The origin of mating system variability and behavioral and demographic correlates of the mating system in Eastern Bluebirds (*Sialia sialis*). Unpub. Ph.D. dissertation, Clemson University, Clemson, South Carolina.

GOWATY, P. A. 1981a. An extension of the Orians-Verner-Willson model to account for mating systems besides polygyny. Am. Nat. 118:851–859.

GOWATY, P. A. 1981b. The aggression of breeding Eastern Bluebirds (*Sialia sialis*) toward each other and intra- and inter-specific intruders. Anim. Behav. 29:1013–1027.

GOWATY, P. A. 1983. Male parental care and apparent monogamy among Eastern Bluebirds (*Sialia sialis*). Am. Nat. 121:149–157.

GOWATY, P. A. 1984. Cuckoldry: the limited scientific usefulness of a colloquial term. Anim. Behav. 32:924–925.

GOWATY, P. A., AND A. A. KARLIN. 1984. Multiple paternity and maternity in single broods of apparently monogamous Eastern Bluebirds. Behav. Ecol. Sociobiol. 15:91–95.

GRANT, P. R. 1972. Convergent and divergent character displacement. Biol. J. Linn. Soc. 4:39–68.

GRANT, P. R. 1975. The classical use of character displacement. Evol. Biol. 2:237–337.

GRANT, P. R., AND T. D. PRICE. 1981. Population variation in continuously varying traits as an ecological genetics problem. Amer. Zool. 21:795–811.

GRAUL, W. D., S. R. DERRICKSON, AND D. W. MOCK. 1977. The evolution of avian polyandry. Am. Nat. 111:812–816.

HAHN, D. C. 1981. Asynchronous hatching in the Laughing Gull: cutting losses and reducing rivalry. Anim. Behav. 29:421–427.

HALL, M. F. 1962. Evolutionary aspects of estrildid song. Symp. Zool. Soc. Lond. 8:37–55.

HALLIDAY, T. R. 1978. Sex selection and mate choice. Pp. 180–213, *In* J. R. Krebs and N. B. Davies (eds.), Behavioural Ecology: An Evolutionary Approach. Sinauer Assoc., Sunderland, Massachusetts.

HALLIDAY, T. R. 1983. The study of mate choice. Pp. 3–32, *In* P. Bateson (ed.), Mate Choice. Cambridge University Press, Cambridge, England.

HAMILTON, W. D. 1964. The genetical evolution of social behaviour. I. and II. J. Theor. Biol. 7:1–52.

HAMILTON, W. J., III., AND G. H. ORIANS. 1965. Evolution of brood parasitism in birds. Condor 67:361–382.

HAMMELL, G. S. 1973. The ecology of Lesser Scaup (*Aythya affinis* Eyton) in southwest Manitoba. Unpub. M.S. thesis, University of Guelph, Guelph, Ontario, Canada.

HANNON, S. J. 1984. Factors limiting polygyny in the Willow Ptarmigan. Anim. Behav. 32:153–161.

HARRIS, G. W. 1970. Hormonal differentiation of the developing central nervous system with respect to patterns of endocrine function. Phil. Trans. R. Soc., Ser. B, 259:165–177.

HAYES, N. F. 1981. The Aviculture, Re-establishment and Status of New Zealand Brown Teal (*Anas aucklandica chlorotis*). Ducks Unlimited (NZ) Inc., Hamilton, New Zealand.

HEINROTH, O. 1911. Beitrage zur Biologie, namentlich Ethologie and Psychologie der Anatiden. Verh 5. Int. Orn. Kong., pp. 589–702.

HOCHBAUM, H. A. 1942. Sex and age determination of waterfowl by cloacal examination. Trans. N. Am. Wildl. Conf. 7:299–307.

HOCHBAUM, H. A. 1944. The Canvasback on a Prairie Marsh. Am. Wildl. Inst., Washington, D.C.

HOGAN-WARBURG, A. J. 1966. Social behaviour of the Ruff *Philomachus pugnax* (L.). Ardea 54:109–229.

HÖHN, E. O. 1943. Some observations on the Common Pochard. British Birds 37:102–107.

HÖHN, E. O. 1947. Sexual behaviour and seasonal changes in the gonads and adrenals of the Mallard. Proc. Zool. Soc. Lond. 117:281–304.

HOWARD, R. D. 1979. Estimating reproductive success in natural populations. Am. Nat. 114:221–231.

HOWE, H. F., AND G. ESTABROOK. 1977. On intraspecific competition for avian dispersers in tropical trees. Am. Nat. 111:817–832.

HOWE, H. F., AND J. SMALLWOOD. 1982. Ecology of seed dispersal. Ann. Rev. Ecol. Syst. 13:210–228.

HUMBURG, D. H., H. H. PRINCE, AND R. A. BISHOP. 1978. The social organization of a Mallard population in northern Iowa. J. Wildl. Manage. 42:72–80.

HUMPHREY, P. S., AND B. C. LIVEZEY. 1982. Flightlessness in flying steamer ducks. Auk 99:368–372.

IMMELMANN, K. 1965. Australian Finches in Bush and Aviary. Halstead Press, Sydney, Australia.

ISTOCK, C. A. 1978. Fitness variation in a natural population. Pp. 171–190, *In* H. Dingle (ed.), Evolution of Insect Migration and Diapause. Springer-Verlag, New York.

JANSON, C. 1983. Adaptation of fruit morphology to dispersal agents in a neotropical forest. Science 219:187–189.

JANZEN, D. H. 1979. How to be a fig. Ann. Rev. Ecol. Syst. 10:13–51.

JEHL, J. R. 1970. Sexual selection for size differences in two species of sandpipers. Evolution 24: 311–319.

JENNI, D. 1974. Evolution of polyandry in birds. Am. Zool. 14:129–144.

JENNI, D., AND G. COLLIER. 1972. Polyandry in the American Jacana *Jacana spinosa*. Auk 89:743–765.

JOHNSGARD, P. A. 1960. Pair-formation in *Anas* (Anatidae) and related genera. Ibis 102:616–618.

JOHNSGARD, P. A. 1965. Handbook of Waterfowl Behavior. Cornell University Press, Ithaca, New York.

JOHNSGARD, P. A. 1968. Waterfowl: their Biology and Natural History. University of Nebraska Press, Lincoln, Nebraska.

JOHNSGARD, P. A. 1978. Ducks, Geese and Swans of the World. University of Nebraska Press, Lincoln, Nebraska.

JOHNSON, A. W. 1965. The Birds of Chile, and Adjacent Regions of Argentina, Bolivia and Peru. Vol. 1. Platt Establecemientos Graficos, Buenos Aires, Argentina.

JOHNSON, D. H., AND A. B. SARGEANT. 1977. Impact of Red Fox predation on the sex ratio of prairie Mallards. U.S. Fish and Wildl. Serv. Wildl. Res. Rept. No. 6. Washington, D.C.

JOHNSON, L. K. 1982. Sexual selection in a brentid weevil. Evolution 36:251–262.

JOHNSON, O. W. 1961. Reproductive cycle of the Mallard duck. Condor 63:251–364.

JOHNSON, O. W. 1966. Quantitative features of spermatogenesis in the Mallard (*Anas platyrhynchos*). Auk 83:233–239.

KANESHIRO, K. Y. 1976. Ethological isolation and phylogeny in the planitibia subgroup of the Hawaiian *Drosophila*. Evolution 30:739–745.

KANESHIRO, K. Y. 1980. Sexual isolation, speciation, and the direction of evolution. Evolution 34: 437–444.

KEAR, J. 1970. The adaptive radiation of parental care in waterfowl. Pp. 357–392, *In* J. H. Crook (ed.), Social Behaviour in Birds and Mammals. Academic Press, New York.

KEAR, J. 1975. Salvadori's Duck of New Guinea. Wildfowl Trust Ann. Rep. 26:104–111.

KEAST, J. A. 1958. Intraspecific variation in the Australian finches. Emu 58:219–246.

KETTERSON, E. D. 1979. Status signaling in Dark-eyed Juncos. Auk 96:94–99.

KIEL, W. H., A. S. HAWKINS, AND N. G. PERRET. 1972. Waterfowl habitat trends in the aspen parkland of Manitoba. Can. Wildl. Rept. Ser. 18.

KIRKPATRICK, M. 1982. Sexual selection and the evolution of female choice. Evolution 36:1–12.

KLEIMAN, D. G. 1977. Monogamy in mammals. Quart. Rev. Biol. 52:39–69.

KLINT, T. 1980. Influence of male nuptial plumage on mate selection in the female Mallard (*Anas platyrhynchos*). Anim. Behav. 28:1230–1238.

KLINT, T., AND M. ENQUIST. 1981. Pair formation and reproductive output in domestic pigeons. Behav. Proc. 6:57–62.

KOENIG, W. D., R. L. MUMME, AND F. A. PITELKA. 1984. The breeding system of the Acorn Woodpecker in central coastal California. Z. Tierpsychol. 65:289–308.

KRAPU, G. L. 1979. Nutrition of female dabbling ducks during reproduction. Pp. 59–70, *In* T. A. Bookhout (ed.), Waterfowl and Wetlands—An Integrated Review. La Crosse Printing Co., La Crosse, Wisconsin.

KRUIJT, J. P., I. BOSSEMA, AND G. J. LAMMERS. 1982. Effects of early experience and male activity on mate choice in Mallard females (*Anas platyrhynchos*). Behaviour 80:32–43.

KUSHLAN, J. A. 1973. Promiscuous mating behavior in the White Ibis. Wilson Bull. 85:331–332.

LACK, D. 1943. The Life of the Robin. H. F. & G. Witherby, London, England.

LACK, D. 1968. Ecological Adaptations for Breeding in Birds. Methuen, London, England.

LANDE, R. 1978. Evolutionary mechanisms of limb loss in tetrapods. Evolution 32:73–92.

LANDE, R. 1980a. Sexual dimorphism, sexual selection and adaptation in polygenic characters. Evolution 34:292–307.

LANDE, R. 1980b. Microevolution in relation to macroevolution. Paleobiol. 6:233–238.

LANDE, R. 1981a. Models of speciation by sexual selection on polygenic traits. Proc. Natl. Acad. Sci. U.S.A. 78:3721–3725.

LANDE, R. 1981b. The minimum number of genes contributing to quantitative variation between and within populations. Genetics 99:541–553.

LANDE, R. 1982. Rapid origin of sexual isolation and character divergence in a cline. Evolution 36: 213–223.

LAURIE-AHLBERG, C. C., AND F. MCKINNEY. 1979. The nod-swim display of male Green-winged Teal (*Anas crecca*). Anim. Behav. 27:165–172.

LEBOEUF, B. J. 1974. Male-male competition and reproductive success in elephant seals. Amer. Zool. 14:163–176.

LEBRET, T. 1961. The pair formation in the annual cycle of the Mallard, *Anas platyrhynchos* L. Ardea 49:97–158.

LENINGTON, S. 1980. Female choice and polygyny in Red-winged Blackbirds. Anim. Behav. 28:347–361.

LEUTHOLD, W. 1966. Variations in territorial behavior of Uganda Kob, *Adenota kob thomasi* (Neuman 1896). Behaviour 26:215–258.

LEVITT, P. R. 1975. General kin selection models for genetic evolution of sib altruism in diploid and haploid species. Proc. Natl. Acad. Sci. U.S.A. 72:4531–4535.

LEWIS, D. M. 1971. Importance of face mask in sexual recognition and territorial behavior in the Yellowthroat. Jack Pine Warbler 50:98–109.

LILL, A. 1974. Evolution of clutch size and male "chauvinism" in the White-bearded Manakin. Living Bird 13:211–231.

LILL, A. 1976. Lek behavior in the Golden-headed Manakin *Pipra erythrocephala* in Trinidad (West Indies). Fort. Verh. 18:1–84.

LORENZ, K. 1950. The comparative method in studying innate behavior patterns. Symp Soc. Exptl. Biol. 4:221–268.

LOW, B. S. 1978. Environmental uncertainty and the parental strategies of marsupials and placentals. Am. Nat. 112:197–213.

LUMPKIN, S. 1981. Avoidance of cuckoldry in birds: the role of the female. Anim. Behav. 29:303–304.

LUMSDEN, C. J., AND E. O. WILSON. 1981. Genes, Mind, and Culture. Harvard University Press, Cambridge, Massachusetts.

MACNAIR, M. R., AND G. A. PARKER. 1978. Models of parent-offspring conflict. II. Promiscuity. Anim. Behav. 26:111–122.

MARLER, P. 1955. Studies of fighting in Chaffinches (II). The effect on dominance relations of disguising females and males. Brit. J. Anim. Behav. 3:137–146.

MATHER, K. 1943. Polygenic inheritance and natural selection. Biol. Rev. 18:32–64.

MAYNARD SMITH, J. 1977. Parental investment: a prospective analysis. Anim. Behav. 25:1–9.

MAYNARD SMITH, J. 1978. The Evolution of Sex. Cambridge University Press, Cambridge, England.

MAYNARD SMITH, J., AND M. G. RIDPATH. 1972. Wife sharing in the Tasmanian Native Hen, *Tribonyx mortierii:* a case of kin selection? Am. Nat. 106:447–452.

MAYR, E. 1968. The sequence of genera in the Estrildidae (Aves). Breviora 287:1–14.

MAYR, E. 1972. Sexual selection and natural selection. Pp. 87–103, *In* B. Campbell (ed.), Sexual Selection and the Descent of Man, 1871–1971. Aldine, Chicago, Illinois.

MAYR, E. 1982. The Growth of Biological Thought. Harvard University Press, Cambridge, Massachusetts.

MCKEY, D. 1975. The ecology of co-evolved seed dispersal systems. Pp. 159–191, *In* L. E. Gilbert and P. H. Raven (eds.), Coevolution of Plants and Animals. University Texas, Austin, Texas.

MCKINNEY, F. 1973. Ecoethological aspects of reproduction. Pp. 6–21, *In* D. S. Farner (ed.), Breeding Biology of Birds. Nat. Acad. Sci., Washington, D.C.

MCKINNEY, F. 1975a. The behaviour of ducks. Pp. 490–519, *In* E. S. E. Hafez (ed.), The Behaviour of Domestic Animals. 3rd ed. Balliere, Tindall and Cassell, London, England.

MCKINNEY, F. 1975b. The evolution of duck displays. Pp. 331–357, *In* G. Baerends, C. Beer, and A. Manning (eds.), Function and Evolution in Behaviour. Clarendon Press, Oxford, England.

MCKINNEY, F. 1986. Ecological factors influencing the social systems of migratory dabbling ducks. Pp. 20, *In* D. I. Rubenstein and R. W. Wrangham (eds.), Ecological Aspects of Social Evolution, Princeton University Press, Princeton. New Jersey.

MCKINNEY, F., J. BARRETT, AND S. R. DERRICKSON. 1978. Rape among Mallards. Science 201:281–282.

MCKINNEY, F., AND D. J. BRUGGERS. 1983. Status and breeding behavior of the Bahama Pintail and the New Zealand Blue Duck. Proc. 1983 Jean Delacour/IFCB Symposium on Breeding Birds in Captivity. Hollywood, California.

MCKINNEY, F., K. M. CHENG, AND D. BRUGGERS. 1984. Sperm competition in apparently monogamous birds. Pp. 523–545, *In* R. L. Smith (ed.), Sperm Competition and the Evolution of Animal Mating Systems. Academic Press, New York.

MCKINNEY, F., S. R. DERRICKSON, AND P. MINEAU. 1983. Forced copulation in waterfowl. Behaviour 86:250–294.

MCKINNEY, F., W. R. SIEGFRIED, I. J. BALL, AND P. G. H. FROST. 1978. Behavioural specializations for river life in the African Black Duck (*Anas sparsa* Eyton). Z. Tierpsychol. 48:349–400.

MCKINNEY, F., AND P. STOLEN. 1982. Extra-pair bond courtship and forced copulation among captive Green-winged Teal (*Anas crecca carolinensis*). Anim. Behav. 30:461–474.

MENDALL, H. L. 1958. The ring-necked duck in the northeast. Univ. Maine Bull. 60:1–317.

MILLS, J. A. 1973. The influence of age and pair-bond on the breeding biology of the Red-billed Gull (*Larus novaehollandiae scopulinus*). J. Anim. Ecol. 42:147–162.

MINEAU, P., AND F. COOKE. 1979. Rape in the Lesser Snow Goose. Behaviour 70:280–291.

MOCK, D. W. 1979. Display repertoire shifts and "extramarital" courtship in herons. Behaviour 69: 57–71.

MOCK, D. W. 1983. On the study of avian mating systems. Pp. 55–84, *In* A. Brush and G. A. Clark, Jr. (eds.), Perspectives in Ornithology. Cambridge University Press, New York.

MOCK, D. W. 1984. Infanticide, siblicide, and avian nestling mortality. Pp. 3–30, *In* G. Hausfater and S. B. Hrdy (eds.), Infanticide: Comparative and Evolutionary Perspectives. Aldine, New York.

MONEY, J., AND A. A. EHRHARDT. 1972. Gender dimorphic behaviour and foetal sex hormones. Recent Prog. Horm. Res. 28:735–763.

MORRIS, D. 1954. The reproductive behaviour of the Zebra Finch (*Poephila guttata*), with special reference to pseudofemale behaviour and displacement activities. Behaviour 6:271–322.

MORRIS, D. 1958. The comparative ethology of grassfinches (Erythrurae) and mannikins (Amadinae). Proc. Zool. Soc. Lond. 131:389–439.

MORTON, E. E., M. S. GEITGEY, AND S. MCGRATH. 1978. On bluebird 'Responses to apparent female adultery.' Am. Nat. 112:968–971.

MULLER, H. J. 1942. Isolating mechanisms, evolution and temperature. Biol. Symp. 6:71–125.

MUMME, R. L., W. D. KOENIG, AND F. A. PITELKA. 1983a. Mate guarding in the Acorn Woodpecker: within-group reproductive competition in a cooperative breeder. Anim. Behav. 31:1094–1106.

MUMME, R. L., W. D. KOENIG, AND F. A. PITEKLA. 1983b. Reproductive competition and communal nesting in the Acorn Woodpecker: sisters destroy each other's eggs. Nature 306:583–584.

MURRAY, B. G., JR. 1979. Population Dynamics: Alternative Models. Academic Press, New York.

MURRAY, B. G., JR. 1984. A demographic theory on the evolution of mating systems as exemplified by birds. Evol. Biol. 18:71–140.

MURRAY, B. G., JR., AND L. GÅRDING. 1984. On the meaning of parameter *x* of Lotka's discrete equations. Oikos 42:323–326.

NELSON, J. B. 1966. The breeding biology of the Gannet, *Sula bassana*, on the Bass Rock, Scotland. Ibis 108:586–626.

NOBLE, G. K. 1936. Courtship and sexual selection in the flicker (*Colaptes auratus luteus*). Auk 53: 269–282.

NOBLE, G. K., AND W. VOGT. 1935. An experimental study of sex recognition in birds. Auk 52:278–286.

NOLAN, V., JR. 1978. Ecology and behavior of the Prairie Warbler *Dendroica discolor*. Ornithol. Monogr. 28:1–595.

NORMAN, F. I. 1982. Eggs, egg-laying and incubation in the Chestnut Teal. Emu 82:195–198.

O'CONNOR, R. J. 1978. Brood reduction in birds: selection for fratricide, infanticide, and suicide? Anim. Behav. 26:79–96.

O'DONALD, P. 1972. Natural selection of reproductive rates and breeding times and its effect on sexual selection. Am. Nat. 106:368–379.

O'DONALD, P. 1973. Frequency-dependent sexual selection as a result of variations in fitness at breeding time. Heredity 30:351–368.

O'DONALD, P. 1974. Polymorphisms maintained by sexual selection in monogamous species of birds. Heredity 32:1–10.

O'DONALD, P. 1977a. Theoretical aspects of sexual selection. Theor. Pop. Biol. 12:298–334.

O'DONALD, P. 1977b. Mating advantage of rare males in models of sexual selection. Nature 267: 151–154.

O'DONALD, P. 1980. Genetic Models of Sexual Selection. Cambridge University Press, New York.

OHDE, B. R., R. A. BISHOP, AND J. J. DINSMORE. 1983. Mallard reproduction in relation to sex ratios. J. Wildl. Manage. 47:118–126.

OHNO, S. 1979. Major sex-determining genes. Monogr. in Endocrin. 11:1–140.

OHTA, A. T. 1978. Ethological isolation and phylogeny in the *grimshawi* species complex of Hawaiian *Drosophila*. Evolution 21:485–492.

OLLASON, J. C., AND G. H. DUNNET. 1978. Age, experience and other factors affecting the breeding success of the Fulmar, *Fulmaris glacialis*, in Orkney. J. Anim. Ecol. 47:961–976.

OLSON, D. P. 1964. A study of canvasback and redhead breeding populations, nesting habitats and productivity. Unpub. Ph.D. dissertation, University of Minnesota, St. Paul, Minnesota.

ORIANS, G. H. 1961. The ecology of blackbird (*Agelaius*) social systems. Ecol. Monogr. 31:285–312.

ORIANS, G. H. 1969. On the evolution of mating systems in birds and mammals. Am. Nat. 103: 589–603.

ORIANS, G. H. 1980. Some adaptations of marsh-nesting blackbirds. Princeton University Press, Princeton, New Jersey.

ORING, L. W. 1982. Avian mating systems. Pp. 1–92, *In* D. S. Farner, J. S. King and K. C. Parkes (eds.), Avian Biology. Vol. 6. Academic Press, New York.

ORING, L. W., AND S. J. MAXSON. 1978. Instances of simultaneous polyandry by a Spotted Sandpiper, *Actitis macularia.* Ibis 120:349–353.

OTTE, D. 1974. Effects and functions in the evolution of signalling systems. Ann. Rev. Ecol. Syst. 5:385–417.

PALMER, R. S. (ED.). 1976. Handbook of North American birds. Vol. 2. Yale University Press, New Haven, Connecticut.

PARKER, G. A. 1979. Sexual selection and sexual conflict. Pp. 123–166, *In* M. S. Blum and N. A. Blum (eds.), Sexual Selection and Reproductive Competition in Insects. Academic Press, New York.

PARKER, G. A., R. R. BAKER, AND V. G. F. SMITH. 1972. The origin and evolution of gamete dimorphism and the male-female phenomenon. J. Theor. Biol. 36:529–553.

PARKER, G. A., AND M. R. MACNAIR. 1978. Models of parent-offspring conflict. I. Monogamy. Anim. Behav. 26:97–110.

PARSONS, P. A. 1980. Isofemale strains and evolutionary strategies in natural poulations. Evol. Biol. 13:175–217.

PARTRIDGE, L. 1980. Mate choice increases a component of offspring fitness in fruit flies. Nature 283:290–291.

PATERSON, H. E. H. 1978. More evidence against speciation by reinforcement. South African J. Science 74:369–371.

PAYNE, R. B. 1977. The ecology of brood parasitism in birds. Ann. Rev. Ecol. Syst. 8:1–28.

PAYNE, R. B. 1979. Sexual selection and intersexual differences in variance of breeding success. Am. Nat. 114:447–452.

PERRINS, C. M. 1970. The timing of birds' breeding seasons. Ibis 112:242–255.

PETERSON, R. T. 1961. A Field Guide to Western Birds. Houghton Mifflin, Boston, Massachusetts.

PICMAN, J. 1977. Intraspecific nest destruction in the Long-billed Marsh Wren (*Telmatodytes palustris palustris*). Can. J. Zool. 55:1997–2003.

PITELKA, F. A., R. T. HOLMES, AND S. F. MACLEAN, JR. 1974. Ecology and evolution of social organization in arctic sandpipers. Amer. Zool. 14:185–204.

POST, W., AND J. S. GREENLAW. 1982. Comparative costs of promiscuity and monogamy: a test of reproductive effort theory. Behav. Ecol. Sociobiol. 10:101–107.

POSTON, H. J. 1969. Relationship between the Shoveler and its breeding habitat at Strathmore, Alberta. Saskatoon Wetlands Seminar. Can. Wildl. Serv. Rep. Series 6:132–137.

POWELL, J. R. 1978. The founder-flush speciation theory: an experimental approach. Evolution 32: 465–474.

POWER, H. W., 1984. Why 'kleptogamy' is not a substitute for 'cuckoldry.' Anim. Behav. 32:923–924.

POWER, H. W., AND C. G. P. DONER. 1980. Experiments on cuckoldry in the Mountain Bluebird. Am. Nat. 116:698–704.

POWER, H. W., E. LITOVICH, AND M. P. LOMBARDO. 1981. Male starlings delay incubation to avoid being cuckolded. Auk 98:386–389.

PRATT, T. K. 1983. Seed dispersal in a montane forest in Papua New Guinea. Unpub. Ph.D. dissertation, Rutgers University, New Brunswick, New Jersey.

QUADAGNO, D. M., R. BRISCOW, AND J. S. QUADAGNO. 1977. Effect of perinatal gonadal hormones on selected nonsexual behavior patterns: a critical assessment of the nonhuman and human literature. Psychol. Bull. 84:62–80.

RAIKOW, R. J. 1973. Locomotor mechanisms in North American ducks. Wilson Bull. 85:295–307.

RAITASUO, K. 1964. Social behaviour of the Mallard, *Anas platyrhynchos,* in the course of the annual cycle. Finn. Paper on Game Res. 24:1–72.

RALLS, K. 1976. Mammals in which females are larger than males. Quart. Rev. Biol. 51:245–276.

RALLS, K. 1977. Sexual dimorphism in mammals: avian models and unanswered questions. Am. Nat. 111:917–938.

REICHHOLF, J. 1964. Ein fall von polygynie bei der Reiherente. Anz. Orn. Ges. Bayern 7:339–340.

RHIJN, J. G. VAN. 1973. Behavioral dimorphism in male ruffs, *Philomachus pugnax*. Behaviour 47: 153–229.

RICHDALE, L. E. 1951. Sexual Behavior in Penguins. University of Kansas Press, Lawrence, Kansas.

RICKLEFS, R. E. 1965. Brood-reduction in the Curve-billed Thrasher. Condor 67:505–510.

RICKLEFS, R. E. 1980. Commentary. Condor 82:476–477.

RINGO, J. M. 1977. Why 300 species of Hawaiian *Drosophila*? The sexual selection hypothesis. Evolution 31:694–696.

ROGERS, J. P. 1959. Low water and Lesser Scaup reproduction near Erickson, Manitoba. Trans. N. Am. Wildl. Conf. 24:216–223.

ROGERS, J. P. 1964. Effect of drought on reproduction of the Lesser Scaup. J. Wildl. Manage. 28: 213–222.

ROHWER, S. 1977. Status signalling in Harris' Sparrows: some experiments in deception. Behaviour 61:107–129.

ROHWER, S., AND P. W. EWALD. 1981. The cost of dominance and advantage of subordination in a badge signaling system. Evolution 35:441–454.

RUBENSTEIN, D. I. 1980. On the evolution of alternative mating strategies. Pp. 65–100, *In* J. E. R. Staddon (ed.), Limits to Action: The Allocation of Individual Behavior. Academic Press, New York.

RUDEGEAIR, T. J. 1975. The reproductive behavior and ecology of the White Ibis. Unpub. Ph.D. dissertation, University of Florida, Gainesville, Florida.

SALT, G. W. 1979. A comment on the use of the term *emergent properties*. Am. Nat. 113:145–148.

SARGEANT, A. B. 1972. Red fox spatial characteristics in relation to waterfowl predation. J. Wildl. Manage. 36:225–236.

SAYLER, R. D., AND A. D. AFTON. 1981. Ecological aspects of Common Goldeneyes *Bucephala clangula* wintering on the upper Mississippi River. Ornis Scand. 12:99–109.

SCHODDE, R. 1976. Evolution of the birds-of-paradise and bowerbirds: a resynthesis. Proc. Int. Ornithol. Congr. 16:137–149.

SCOTT, D. K. 1980. Functional aspects of the pair bond in winter Bewick's Swans (*Cygnus columbianus bewickii*). Behav. Ecol. Sociobiol. 7:323–327.

SEARCY, W. A. 1979a. Female choice of mates: a general model for birds and its application to Red-winged Blackbirds (*Agelaius phoeniceus*). Am. Nat. 114:77–100.

SEARCY, W. A. 1979b. Male characteristics and pairing success in Red-winged Blackbirds. Auk 96: 353–363.

SEEL, D. C. 1968. Clutch size, incubation and hatching success in the House Sparrow and Tree Sparrow, *Passer* spp., at Oxford. Ibis 110:270–282.

SELANDER, R. K. 1972. Sexual selection and dimorphism in birds. Pp. 180–230, *In* B. Campbell (ed.), Sexual Selection and the Descent of Man, 1871–1971. Aldine, Chicago, Illinois.

SEMLER, D. E. 1971. Some aspects of adaptation in a polymorphism for breeding colours in the threespine stickleback (*Gasterosteus aculeatus*). J. Zool. (Lond.) 165:291–302.

SERIE, J. R., D. L. TRAUGER, H. A. DOTY, AND D. E. SHARP. 1982. Age-class determination of Canvasbacks. J. Wildl. Manage. 46:894–904.

SEYMOUR, N. R. 1974. Territorial behaviour of wild Shovelers at Delta, Manitoba. Wildfowl 25: 49–55.

SEYMOUR, N. R., AND R. D. TITMAN. 1978. Changes in activity patterns, agonistic behavior, and territoriality of Black Ducks (*Anas rubripes*) during the breeding season in a Nova Scotia tidal marsh. Can. J. Zool. 56:1773–1785.

SHIELDS, W. M. 1977. The social significance of avian winter plumage variability: a comment. Evolution 31:905–907.

SHIELDS, W. M. 1982. Philopatry, Inbreeding, and the Evolution of Sex. SUNY, Albany, New York.

SHEPARD, J. M. 1975. Factors influencing female choice in the lek mating system of the Ruff. Living Bird 14:87–111.

SHERMAN, P. W. 1981. Electrophoresis and avian genealogical analyses. Auk 98:419–421.

SIBLEY, C. G. 1957. The evolutionary and taxonomic significance of sexual dimorphism and hybridization in birds. Condor 59:166–191.

SIEGEL, S. 1956. Nonparametric Statistics. McGraw-Hill, New York.

SIEGFRIED, W. R. 1974. Brood care, pair bonds and plumage in southern African Anatini. Wildfowl 25:33–40.

SIEGFRIED, W. R. 1979. Social behavior of the African Comb Duck. Living Bird 17:85–104.

SIEGFRIED, W. R., P. G. H. FROST, AND C. W. HEYL. 1976. Long-standing pair-bonds in Cape Teal. Ostrich 47:130–131.

SKUTCH, A. F. 1935. Helpers at the nest. Auk 52:257–273.

SKUTCH, A. F. 1976. Parent Birds and Their Young. University of Texas Press, Austin, Texas.

SMITH, D. G. 1972. The role of epaulets in the Red winged Blackbird (*Agelaius phoeniceus*) social system. Behaviour 41:251–268.

SMITH, N. G. 1966. Evolution of some arctic gulls (*Larus*): an experimental study of isolating mechanisms. Ornithol. Monogr. 4:1–99.

SMITH, R. I. 1968. The social aspects of reproductive behavior in the Pintail. Auk 85:381–396.

SMITH, S. M. 1980. Henpecked males: the general pattern in monogamy? J. Field Ornithol. 51: 56–64.

SNOW, B. K. 1973. Notes on the behavior of the White Bellbird. Auk 90:743–751.

SNOW, B. K., AND D. W. SNOW. 1979. The Ochre-bellied Flycatcher and the evolution of lek behavior. Condor 81:286–645.

SNOW, D. W. 1962. A field study of the Black and White Manakin, *Manacus manacus*, in Trinidad. Zoologica 47:183–198.

SNOW, D. W. 1976. The Web of Adaptation. Quadrangle Press, New York.

SNOW, D. W. 1980. Regional differences between tropical floras and the evolution of frugivory. Proc. Int. 17th Ornithol. Congr., pp. 1192–1198.

SOKAL, R. R., AND F. J. ROHLF. 1981. Biometry. 2nd ed. W. H. Freeman and Co., San Francisco, California.

SOWLS, L. K. 1955. Prairie Ducks. Wildl. Mgmt. Inst., Washington, D.C.

SPIETH, H. T. 1974. Mating behavior and evolution of the Hawaiian *Drosophila*. Pp. 94–110, *In* M. J. D. White (ed.), Genetic Mechanisms of Speciation in Insects. Australia and New Zealand Book Company, Sydney, Australia.

STACEY, P. B., AND T. C. EDWARDS, JR. 1983. Possible cases of infanticide resulting from competition for reproductive status in a group-breeding bird. Auk 100:731–733.

STANDEN, P. J. 1976. The social behaviour of the Chilean Teal. Unpub. Ph.D. thesis, University of Leicester, Leicester, England.

STANDEN, P. J. 1980. The social display of the Chilean Teal *Anas flavirostris flavirostris*. J. Zool., Lond. 191:293–313.

STEELE, E. J. 1981. Somatic Selection and Adaptive Evolution: on the Inheritance of Acquired Characters. 2nd ed. University of Chicago Press, Chicago, Illinois.

STEINER, H. 1960. Die Klassifikation der Prachtfinken, Spermestidae, auf Grund der Rachenzeichungen ihrer Nestlinge. J. Ornith. 101:92–112.

STEWART, G. R., AND R. D. TITMAN. 1980. Territorial behaviour by prairie pothole Blue-winged Teal. Can. J. Zool. 58:639–649.

STINSON, C. H. 1979. On the selective advantage of fratricide in raptors. Evolution 33:1219–1225.

STOLEN, P., AND F. MCKINNEY. 1983. Bigamous behaviour of Cape Teal. Wildfowl 34:10–13.

STOUDT, J. H. 1965. Project report on habitat requirements of Canvasback during the breeding season. U.S. Fish and Wildl. Serv. Proj. A-8.

STOUDT, J. H. 1971. Ecological factors affecting waterfowl production in Saskatchewan parklands. U.S. Fish and Wildl. Serv. Resour. Pub. 99.

STOUDT, J. H. 1982. Habitat use and productivity of Canvasbacks in southwestern Manitoba, 1961–1972. U.S. Fish and Wildl. Serv., Spec. Sci. Rept., Wildl. No. 248.

SUGDEN, L. G., AND H. J. POSTON. 1968. A nasal marker for ducks. J. Wildl. Manage. 32:984–986.

TEMPLETON, A. R. 1979. Once again, why 300 species of Hawaiian *Drosophila*? Evolution 33:513–517.

THOMPSON, D. W. 1942. On Growth and Form. Cambridge University Press, Cambridge, England.

THORNHILL, R. 1976. Sexual selection and nuptial feeding behavior in *Bittacus apicalis* (Insecta: Mecoptera). Am. Nat. 110:529–548.

THORNHILL, R. 1980a. Rape in *Panorpa* scorpionflies and a general rape hypothesis. Anim. Behav. 28:52–59.

THORNHILL, R. 1980b. Mate choice in *Hylobittacus apicalis* (Insecta: Mecoptera) and its relation to some models of female choice. Evolution 34:519–538.

THORNHILL, R., AND J. ALCOCK. 1983. The Evolution of Insect Mating Systems. Harvard University Press, Cambridge, Massachusetts.

TINBERGEN, N. 1942. An objective study of the innate behaviour of animals. Biblioth. Biotheor. 1:39–98.

TITMAN, R. D. 1981. A time-activity budget for breeding Mallards (*Anas platyrhynchos*) in Manitoba. Canadian Field-Nat. 95:266–271.

TITMAN, R. D., AND N. R. SEYMOUR. 1981. A comparison of pursuit flights by six North American ducks of the genus *Anas*. Wildfowl 32:11–18.

TODD, F. S. 1979. Waterfowl: Ducks, Geese, and Swans of the World. Sea World Press, San Diego, California.

TRAIL, P. W., S. D. STRAHL, AND J. L. BROWN. 1981. Infanticide in relation to individual and flock histories in a communally breeding bird, the Mexican Jay (*Aphelocoma ultramarina*). Am. Nat. 118:72–82.

TRAUGER, D. L., AND J. H. STROUDT. 1978. Trends in waterfowl populations and habitats on study areas in Canadian parklands. Trans. N. Am. Wildl. Natur. Resour. Conf. 43:187–205.

TRIVERS, R. L. 1972. Parental investment and sexual selection. Pp. 136–179, *In* B. Campbell (ed.), Sexual Selection and the Descent of Man, 1871–1971. Aldine, Chicago, Illinois.

TRIVERS, R. L. 1974. Parent-offspring conflict. Amer. Zool. 14:249–264.

VAN RHIJN, J. G. 1973. Behavioural dimorphism in male ruffs, *Philomachus pugnax*. Behaviour 47: 153–229.

VAN RYZIN, M. T., AND H. I. FISHER. 1976. The age of Laysan Albatrosses, *Diomedea immutabilis*, at first breeding. Condor 78:1–9.

VEHRENCAMP, S. L. 1977. Relative fecundity and parental effort in communally nesting anis, *Crotophaga sulcirostris*. Science 197:403–405.

VERNER, J. 1964. Evolution of polygamy in the Long-billed Marsh Wren. Evolution 18:252–261.

VERNER, J. 1977. On the adaptive significance of territoriality. Am. Nat. 111:769–775.

VERNER, J., AND M. F. WILLSON. 1966. The influence of habitats on mating systems of North American passerine birds. Ecology 47:143–147.

VERNER, J., AND M. F. WILLSON. 1969. Mating systems, sexual dimorphism, and the role of male North American Passerine birds in the nesting cycle. Ornithol. Monogr. 9:1–76.

WADDINGTON, C. H. 1957. The Strategy of the Gene. Allen and Unwin, London, England.

WADE, M. J. 1979. The evolution of social interactions by family selection. Am. Nat. 113:399–417.

WADE, M. J., AND S. J. ARNOLD. 1980. The intensity of sexual selection in relation to male sexual behaviour, female choice, and sperm precedence. Anim. Behav. 28:446–461.

WALL, VON DE, W. 1965. "Gesellschaftsspiel" und Balz der Anatini. J. Ornithol. 106:65–80.

WARD, P., AND B. D. J. BATT. 1973. Propagation of Captive Waterfowl. Wildl. Mgmt. Inst., Washington, D.C.

WASSER, S. K. (ED.). 1983. Social Behavior of Female Vertebrates. Academic Press, New York.

WEATHERHEAD, P. J., AND R. J. ROBERTSON. 1978. Intraspecific nest parasitism in the Savannah Sparrow. Auk 95:744–745.

WEIDMANN, U. 1956. Verhaltensstudien an der Stockente (*Anas platyrhynchos* L.) I. Das Aktionssystem. Z. Tierpsychol. 13:208–271.

WEIDMANN, U., AND J. DARLEY. 1971. The role of the female in the social display of Mallards. Anim. Behav. 19:287–298.

WELLER, M. W. 1957. An automatic nest trap for waterfowl. J. Wildl. Manage. 21:456–458.

WELLER, M. W. 1959. Parasitic egg laying in the Redhead (*Aythya americana*) and other North American Anatidae. Ecol. Monogr. 29:333–365.

WELLER, M. W. 1964. The reproductive cycle. Pp. 35–79, *In* J. Delacour (ed.), The Waterfowl of the World. Country Life, London, England.

WELLER, M. W. 1965. Chronology of pair formation in some nearctic *Aythya* (Anatidae). Auk 82: 227–235.

WELLER, M. W. 1967. Notes on some marsh birds of Cape San Antonio, Argentina. Ibis 109:391–416.

WELLER, M. W. 1968. Notes on some Argentine anatids. Wilson Bull. 80:189–212.

WELLER, M. W. 1972. Ecological studies of Falkland Islands' waterfowl. Wildfowl 23:25–44.

WELLER, M. W. 1975a. Habitat selection by waterfowl of Argentine Isla Grande. Wilson Bull. 87: 83–90.

WELLER, M. W. 1975b. Ecological studies of the Auckland Islands Flightless Teal. Auk 92:280–297.

WELLER, M. W. 1975c. Ecology and behaviour of the South Georgia Pintail *Anas g. georgica.* Ibis 117:217–231.

WELLER, M. W. 1980. The Island Waterfowl. Iowa State University Press, Ames, Iowa.

WELLS, K. D. 1977. The social behaviour of anuran amphibians. Anim. Behav. 25:666–693.

WICKLER, W., AND U. SEIBT. 1983. Monogamy: an ambiguous concept. Pp. 33–52, *In* P. Bateson (ed.), Mate Choice. Cambridge University Press, Cambridge, England.

WIEBES, J. T. 1979. Co-evolution of figs and the insect pollinators. Ann. Rev. Ecol. Syst. 10:1–12.

WILEY, R. H. 1973. Territoriality and non-random mating in Sage Grouse, *Centrocercus urophasianus.* Anim. Behav. Monogr. 6:85–169.

WILLIAMS, D. M. 1983. Mate choice in the Mallard. Pp. 297–309, *In* P. Bateson (ed.), Mate Choice. Cambridge University Press, Cambridge, England.

WILLIAMS, G. C. 1966. Adaptation and Natural Selection. Princeton University Press, Princeton, New Jersey.

WILLIAMS, G. C. 1975. Sex and Evolution. Princeton University Press, Princeton, New Jersey.

WILLIAMS, L. 1952. Breeding behavior of the Brewer Blackbird. Condor 54:3–47.

WILLSON, M. F., AND N. BURLEY. 1983. Mate choice in plants: tactics, mechanisms, and consequences. Princeton University Press, Princeton. New Jersey.

WILSON, E. O. 1975. Sociobiology, the new Synthesis. Harvard University Press, Cambridge, Massachusetts.

WINTERBOTTOM, J. M. 1974. The Cape Teal. Ostrich 45:110–132.

WISHART, R. A. 1983. Pairing chronology and mate selection in the American Wigeon (*Anas americana*). Can. J. Zool. 61:1733–1743.

WISHART, R. A., AND R. W. KNAPTON. 1978. Male Pintails defending females from rape. Auk 95: 186–187.

WITTENBERGER, J. F. 1976. The ecological factors selecting for polygyny in altricial birds. Am. Nat. 110:779–799.

WITTENBERGER, J. F. 1979. The evolution of mating systems in birds and mammals. Pp. 271–349, *In* P. Marler and J. Vandenbergh (eds.), Handbook of Behavioral Neurobiology. Vol. 3. Plenum Press, New York.

WITTENBERGER, J. F. 1981. Animal Social Behavior. Duxbury Press, Boston, Massachusetts.

WITTENBERGER, J. F., AND R. L. TILSON. 1980. The evolution of monogamy: hypotheses and evidence. Ann. Rev. Ecol. Syst. 11:197–232.

WRIGHT, S. 1977. Evolution and the Genetics of Populations. Vol. III. Experimental results and evolutionary deductions. University of Chicago Press, Chicago, Illinois.

WUNDERLE, J. M., JR. 1984. Mate switching and a seasonal increase in polygyny in the Bananaquit. Behaviour 88:123–144.

YOM-TOV, Y. 1974. The effect of food and predation on breeding density and success, clutch size and laying date of the Crow (*Corvus corone* L.). J. Anim. Ecol. 43:470–498.

YOM-TOV, Y. 1980. Intraspecific nest parasitism in birds. Biol. Rev. 55:93–108.

YOM-TOV, Y., G. M. DUNNETT, AND A. ANDERSSON. 1974. Intraspecific nest parasitism in the Starling *Sturnus vulgaris.* Ibis 116:87–90.

ZAHAVI, A. 1975. Mate selection—a selection for a handicap. J. Theor. Biol. 53:205–214.

ZAHAVI, A. 1977. Reliability in communication systems and the evolution of altruism. Pp. 253–259, *In* B. Stonehouse and C. Perrins (eds.), Evolutionary Ecology. University Park Press, Baltimore, Maryland.

ZENONE, P. G., M. E. SIMS, AND C. J. ERICKSON. 1979. Male Ring Dove behavior and the defense of genetic paternity. Am. Nat. 114:615–626.

ZIMMERMAN, J. L. 1966. Polygyny in the Dickcissel. Auk 83:534–546.